Business Guides on the Go

"Business Guides on the Go" presents cutting-edge insights from practice on particular topics within the fields of business, management, and finance. Written by practitioners and experts in a concise and accessible form the series provides professionals with a general understanding and a first practical approach to latest developments in business strategy, leadership, operations, HR management, innovation and technology management, marketing or digitalization. Students of business administration or management will also benefit from these practical guides for their future occupation/careers.

These Guides suit the needs of today's fast reader.

Josef Baker-Brunnbauer

Trustworthy Artificial Intelligence Implementation

Introduction to the TAII Framework

 Springer

Josef Baker-Brunnbauer (iD)
Leonding, Austria

ISSN 2731-4758   ISSN 2731-4766   (electronic)
Business Guides on the Go
ISBN 978-3-031-18274-7   ISBN 978-3-031-18275-4   (eBook)
https://doi.org/10.1007/978-3-031-18275-4

This Springer imprint is published by the registered company Springer Nature Switzerland AG.
The registered company address is: Gewerbestrasse 11, 6330 Cham, Switzerland

Prelude

This book addresses the management awareness about ethical and moral aspects of Artificial Intelligence (AI). It is a general trend to speak about AI and many start-ups and established companies are communicating about the development and implementation of AI solutions. There are many events and marketing workshops about AI solutions that are often driven by the development of the technology. AI systems have immense potential to change many domains like the education system, legal and law system, retail market, whole industries, and societies. Therefore, it is important to consider different perspectives besides the technology. As data is one key element for AI, data protection and legal implementation will change. The way in which societies are interacting and organizing themselves will be affected. Such changes require a multi-perspective of the humanity for shaping the future. Will AI systems promote the high-developed nations, or will it take care of the poorest? The development of AI systems is not only a technical discipline: it requires an interaction between multi-professions. This book is aiming to overcome those barriers with the results of a fundamental literature and empirical research to answer the question: *What kind of awareness does the management have about the social impact of their AI product or service?*

This question is divided into five sub-questions that will be answered by a fundamental literature and an empirical research study. This covers the management understanding of the terms moral, ethics, and artificial intelligence, the internal company prioritization of moral and ethics, the involved stakeholders in the AI product or service development and it will analyze the known and used ethical AI guidelines and principles. In the end, the social responsibility of the management regarding their AI system is analyzed and compared. This research has not the aim to discuss AI on a technical level, it is analyzing the management awareness about their social impact to shape the future.

Organizations and companies need practical tools and guidelines to kick-off the implementation of Trustworthy Artificial Intelligence (TAI) systems. AI development companies are still in the beginning of this process or have not even started yet. The findings of the research address to decrease the entry level barrier for AI ethics implementation by introducing the Trustworthy Artificial Intelligence Implementation (TAII) Framework. The outcome is comparatively unique given that it considers a meta perspective of implementing TAI within organizations. As such, the framework aims to fill a literature gap for management guidance to tackle trustworthy AI implementation while considering ethical dependencies within the company. The TAII Framework* takes a holistic approach to identify the systemic relationships of ethics for the company ecosystem and considers corporate values, business models, and common good aspects like the Sustainable Development Goals and the Universal Declaration of Human Rights. The TAII Framework creates guidance to initiate the implementation of AI ethics in organizations without requiring a deep background in philosophy and considers the social impacts outside of a software and data engineering setting. Depending on the legal regulation or area of application, the TAII Framework can be adapted and used with different regulations and ethical principles.

Acknowledgement

I highly appreciate the support of all participants and companies in the empirical research study and all people who shared their thoughts, opinions, and time with me. Moreover, I want to thank Prof. Dr. Joanna Bryson, Prof. Dr. Virginia Dignum, and Prof. Dr. Sandra Wachter for inspiration, professional feedback, and motivation. Ethics of Artificial Intelligence is and will be a multi-profession topic that is generating strong impact for societies.

Finally, I also want to thank my wife Nathalie for her support and understanding of being short of time.

Contents

About the Author

 **Josef Baker-Brunnbauer** worked for over 20 years in different industry areas for start-ups and established companies in international projects for market leading clients. He is consulting companies about business model innovation, product innovation, change and digital transformation projects through his consulting company product-xyz.com and leading research projects for SocialTechLab.eu. The research focus is on socio-technical systems, trustworthy artificial intelligence, and its social impact. Besides, he is a member of the AI Alliance at the European Commission.He holds a master's degree in international business administration (MBA) from LIMAK Johannes Kepler University Linz/Austria and Tsinghua University Beijing/China, an engineering degree in software engineering (Ing), a diploma in life counseling and coaching (LSB), in enterprise design thinking from IBM, in product management (FH) from Marketing Academy Munich and University of Applied Sciences Upper Austria, and a master's degree for psychosocial counseling (MSc) from the Karl-

Franzens University in Graz/Austria.In his leisure time, he likes to travel, discover new opportunities outside his comfort zone, cooking, and endurance sports. Thoughts and feedback are welcome: josef@baker-brunnbauer.com

Abbreviations

AAAI	Association for the Advancement of Artificial Intelligence
AI	Artificial Intelligence
AI HLEG	Artificial Intelligence High-Level Expert Group
AGI	Artificial General Intelligence
AWS	Autonomous Weapon System
BM	Business Model
BMI	Business Model Innovation
CE	Conformité Européenne
CHAR	Character
DL	Deep Learning
DNA	Deoxyribonucleic Acid
EC	European Commission
EI	Emotional Intelligence
FDA	Food and Drug Administration
FLOPS	Floating-point operations per second
fMRI	Functional Magnetic Resonance Imaging
FRA	The European Union Agency for Fundamental Rights
GDPR	General Data Protection Regulation
GOFAI	Good Old-Fashioned AI
GPT-3	Generative Pretrained Transformer 3
ISS	International Space Station
IT	Information Technology
IoT	Internet of Things

MIT	Massachusetts Institute of Technology
ML	Machine Learning
MRI	Magnetic Resonance Imaging
MVEP	Minimum Viable Ethical Product
NASA	National Aeronautics and Space Administration
SME	Small and Medium-size Enterprises
TAI	Trustworthy Artificial Intelligence
TAII	Trustworthy Artificial Intelligence Implementation
TÜV	Technischer Überwachungsverein

List of Figures

List of Tables

1

Introduction

This book aims to generate awareness of ethical challenges for Artificial Intelligence (AI) systems and to analyze the management perspective and understanding of ethics for their AI product or service. While analyzing existing ethical standards and principles for AI systems, the work will generate a better understanding and improvement of ethical principles and moral values for the AI era. This work will contribute to research in the field of ethical standards for AI systems and will represent the management perspective (C-level, executives, etc.). The work will also be a beneficial contribution to generating and implementing future ethical AI guidelines and procedures. Working on ethical and moral understanding regarding AI systems and especially the implementation, is still in an early stage. It is not only about defining what is right and what is wrong as it is such a complex work that can also not be done by a small group of professionals. But it is timely and important to discuss, agree, and standardize common principles.

As digitalization covers many different technologies and aspects, AI can be seen as one of them that will not only change businesses but humanity as well. Besides new technologies and use cases, AI will have a deep impact on society, social life, and has the potential to seriously shape

© The Author(s), under exclusive license to Springer Nature Switzerland AG 2023
J. Baker-Brunnbauer, *Trustworthy Artificial Intelligence Implementation*, Business
Guides on the Go, https://doi.org/10.1007/978-3-031-18275-4_1

and change Digital Humanism. An increasing digitalization and digitization on all levels not only create an improvement and optimization of the products and processes, but it also changes the way of internal and external collaborations (Matzler et al., 2016). A high level of automation with AI systems generates not only a good rating of the company's performance as it also offers the potential to eliminate existing jobs and increases the psychological pressure on employees. Digitalization is not only a challenge to employees executing simple tasks with a low job education, but also to middle and high management.

One of the current trending technology topics is the development and implementation of Artificial Intelligence (AI). Technology market leaders like Google have been long investing billions of money in innovation, research, and implementation of AI alongside trying to retain the image of a search engine company (CBInsights, n.d.). Meanwhile, AI reached the mainstream tech market and established companies, as well as start-ups, are working on AI products and services. Besides the economical aspect, AI has been discovered as a future weapon by politicians to influence the world order (Lee, 2018). Especially China and the USA are investing billions of money in new technologies. Compared to the number of patents, Europe is already far behind (WIPO, 2019). Management of established companies and start-ups, who are working on AI products and services, are responsible for how the future will look like. From a consumer perspective, companies from the USA like Google, Apple, and IBM or Chinese leaders like Baidu, Alibaba, and Tencent (BAT) are innovating new technology products and services to create a customer experience and offer convenience. But there are other companies like, for example, Boston Dynamics, which is working on new generation robots that can be easily used as war soldiers. Developing such nonhuman products in combination with AI technology is a momentous responsibility for humankind.

1.1   The Challenge

As a first step, a general literature research will clarify terms, common understanding, ethical status, and AI principles. In the second step, the empirical research part will analyze the management perspective regarding AI ethics. Bostrom and Yudkowsky (2014) analyzed the moral status of machines and future AI moral aspects when it will be self-developed by the system. Others (Rothenberger et al., 2019; Floridi et al., 2018) analyzed and created an ethical AI guideline or identified moral issues with AI (Siau & Wang, 2018). Therefore, the motivation for starting the AI development defines the research and development setting. Ethical aspects and social responsibility of the management will be analyzed in the expert interviews.

Digital technologies have important economic and social aspects. Companies strive for product innovations and inventions with new technology. Besides this, the long-term impact of digital transformation and new technologies like AI is not clear from the beginning. Digital technologies, products and services are often developed by computer scientists for technical usage with a focus on revenue and growth. Often, social components, especially considering the big picture of humankind or social responsibility purpose, do not have a high priority for the management. Therefore, in consideration of the circumstances of AI, companies should actively take social responsibility beforehand and reconsider their prioritization. This work will analyze the ethical aspects of AI implementations from the management perspective and actively create awareness within the management about their power of social impact. Therefore, companies will need to increase flexibility and openness to innovate new business models with the intelligent usage of new technology like AI (Baker-Brunnbauer, 2019). As there are fewer existing ethical AI research studies from a management perspective, the focus is to further investigate these challenges by answering the following questions.

What kind of awareness does the management have about the social impact of their Artificial Intelligence (AI) product or service?

Therefore, the focus is on the central European area without any consideration of international cultural influence. The focus is to further investigate these challenges by answering the questions based on the following research field:

- *What kind of definitions for AI, ethics, and moral are used by the management?*
- This question will analyze the understanding of the terms by the management.
- *What kind of prioritization does ethics and moral have for the management?*
- This question will analyze the awareness and importance of ethics and moral.
- *Who are the responsible internal and external stakeholders for designing the AI product or service? What kind of professions does the team have?*
- This question will analyze the diversity of the involved professions.
- *Are there AI guidelines and principles known and used within the company?*
- This question will analyze the usage and content of AI guidelines.
- *What kind of social responsibility does the management mention in a self-reflection?*
- This question will analyze the management's awareness of (potential) social impact.

1.2   Course of Action

This work analyses the management perspective of AI companies from an ethical and moral aspect. Therefore, literature and research papers (Floridi et al., 2018; Pagallo, 2017; Indurkhya, 2019; Ransbotham et al., 2019; Siau & Wang, 2018; Bostrom & Yudkowsky, 2014; Tegmark, 2018; Rothenberger et al., 2019) are analyzed and verified with the expert interview results. The outcome shows, if managers consider ethical and moral aspects and how the setting looks like. The target of the interview questions will be to understand the motivation behind the decision makers and analyze their awareness about their influence on social impact. It will also clarify, if existing ethical AI guidelines or codes of conduct are used and what are the pros and cons of their implementation.

Therefore, analyses of guidelines for AI implementations were done. First, to create a fundamental background and understanding of the current research and literature state, research about methodologies and approaches were done. Second, the empirical analysis started with the definition of the interview guidelines and questions. Based on the outcome of the literature study, main categories have been defined to emphasize the focus on the central research question. Nine interview candidates have been chosen randomly from the pool of AI management.

This book will answer the defined question in Sect. 1.1. Therefore, the document will start to give an extract of the most important terms, as there is no central definition of AI. This book is structured in a fundamental literature and in an empirical research part. The literature research in Chap. 2 describes the most used understandings of AI, the current state, and visible challenges of AI. Followed by Chap. 3, which analyses existing literature and research outcomes related to ethics and moral values. Therefore, the focus is on training, challenges, and guidelines. Chapter 4 describes the perspective of the management regarding AI implementations, strategy, and decisions. Followed by Chap. 5 which describes the used method and research setting for the empirical research, which is done via expert interviews. The findings of the first empirical research are summarized in Chap. 6. Finally, the findings were transformed into further research and for the generation of the TAII Framework described in Chap. 7.

References

Baker-Brunnbauer, J. (2019). *Business model innovation in a paradoxical area of conflict (executive summary)*. https://doi.org/10.13140/RG.2.2.24272.66566

Bostrom, N., & Yudkowsky, E. (2014). The ethics of artificial intelligence. In K. Frankish & W. Ramsey (Eds.), *The Cambridge handbook of artificial intelligence* (pp. 316–334). Cambridge University Press. https://doi.org/10.1017/CBO9781139046855.020

CBInsights. (n.d.). *Alphabet in AI: How Google went from a search engine to an $800B global AI powerhouse*. Accessed October 29, 2019, from https://www.cbinsights.com/research/report/alphabet-google-artificial-intelligence

Floridi, L., Cowls, J., Beltrametti, M., Chatila, R., Chazerand, P., Dignum, V., Luetge, C., Madelin, R., Pagallo, U., Rossi, F., Schafer, B., Valcke, P., & Vayena, E. (2018). AI4People—An ethical framework for a good AI society: Opportunities. *Risks Principles Recommend Minds Mach, 28*, 689–707. https://doi.org/10.1007/s11023-018-9482-5

Indurkhya, B. (2019). Is morality the last frontier for machines? *New Ideas Psychol, 54*, 107–111. ISSN: 0732-118X. https://doi.org/10.1016/j.newideapsych.2018.12.001

Lee, K.-F. (2018). *AI Super-Powers*. Houghton Mifflin Harcourt.

Matzler, K., Bailom, F., Friedrich von den Eichen, S., & Anschober, M. (2016). *Wie Sie Ihr Unternehmen digital auf das digitale Zeitalter Disruption vorbereiten*. Vahlen.

Pagallo, U. (2017). When morals ain't enough: Robots, ethics, and the rules of the law. *Minds & Machines, 27*, 625–638. https://doi.org/10.1007/s11023-017-9418-5

Ransbotham, S., Khodabandeh, S., Fehling, R., LaFountain, B., & Krion, D. (2019). *Winning with AI*. MIT Sloan Management Review and Boston Consulting Group. Accessed November 28, 2019, from https://sloanreview.mit.edu/ai2019

Rothenberger, L., Fabian, B., & Arunov, E. (2019). Relevance of ethical guidelines for Artificial Intelligence – A survey and evaluation. In *Proceedings of the 27th European Conference on Information Systems (ECIS)*, Stockholm & Uppsala, Sweden, June 8–14, 2019. ISBN 978-1-7336325-0-8 Research-in-Progress Papers. https://aisel.aisnet.org/ecis2019_rip/26

Siau, K., & Wang, W. (2018). *Ethical and moral issues with AI - a case study on healthcare robots*. Emergent Research Forum (ERF). Accessed October 20, 2019, from https://www.researchgate.net/publication/325934375_Ethical_and_Moral_Issues_with_AI

Tegmark, M. (2018). *Life 3.0*. Penguin Random House.

WIPO. (2019). *WIPO Technology Trends 2019: Artificial Intelligence*. Accessed October 29, 2019, from https://www.wipo.int/edocs/pubdocs/en/wipo_pub_1055.pdf

2

Artificial Intelligence

2.1   Terms and Definitions

Literature and empirical research work confirmed that people are using the terms "Artificial Intelligence, Moral and Ethics" in different contexts and meanings. As there exists more than one definition and understanding, it is important to clarify the common understanding within research and working groups. To get an understanding of this research, an extract overview of used definitions is listed:

Moral

- The Oxford Learner's Dictionaries describes moral by principles of wrong and right behavior (Oxford Learner's Dictionaries, n.d.).

Ethics

- The Oxford Learner's Dictionaries defines ethics as "moral principles that control or influence a person's behaviour" (Oxford Learner's Dictionaries, n.d.).

© The Author(s), under exclusive license to Springer Nature Switzerland AG 2023
J. Baker-Brunnbauer, *Trustworthy Artificial Intelligence Implementation*, Business Guides on the Go, https://doi.org/10.1007/978-3-031-18275-4_2

- According to MIT professor Max Tegmark, ethics is about "Principles that govern how we should behave" (Tegmark, 2018).
- The European Commission describes ethics as an academic discipline that is a subfield of philosophy. Ethics covers four major fields of research: Meta-ethics, normative ethics, descriptive ethics, and applied ethics. AI ethics is seen as an example of applied ethics and focuses on normative issues generated by design, development, implementation, and usage of AI (European Commission, 2019b).

Intelligence

- The Oxford Learner's Dictionaries uses "the ability to learn, understand and think in a logical way about things" (Oxford Learner's Dictionaries, n.d.).
- According to Tegmark, intelligence defines the ability to accomplish complex goals. As there are different possible goals (physical, social, and technical), there are many different types of intelligence like narrow (e.g., chess computer) and broad intelligence—AGI (Tegmark, 2018).
- Further development of intelligence is Emotional Intelligence (EI): The Oxford Learner's Dictionaries defines EI as "the ability to understand your emotions and those of other people and to behave appropriately in different situations" (Oxford Learner's Dictionaries, n.d.).

Artificial Intelligence

- The Oxford Learner's Dictionaries defines AI as "an area of study concerned with making computers copy intelligent human behaviour" (Oxford Learner's Dictionaries, n.d.).
- AI can be classified into weak AI, which can only process specific tasks and strong AI (AGI), which is able to process multiple tasks with human intelligence. Therefore, researchers are concerned that AGI will lead to superintelligence that exceeds human cognitive performance (Siau & Wang, 2018).

- Researchers define AI via those aspects: Intelligence as a set of higher cognitive abilities which places humans separately from other animals (Rothenberger et al., 2019).
- Others use the definition of AI from the Oxford English Dictionary, which describes it as the theory and development of computer systems that can perform tasks, which normally require human intelligence (Ransbotham et al., 2019).
- The European Commission (2019a) defines AI systems as "[...] software (and possibly also hardware) systems designed by humans that, given a complex goal, act in the physical or digital dimension by perceiving their environment through data acquisition, interpreting the collected structured or unstructured data, reasoning on the knowledge, or processing the information, derived from this data and deciding the best action(s) to take to achieve the given goal. [...] As a scientific discipline, AI includes several approaches and techniques, such as machine learning [...], machine reasoning [...] and robotics [...]." On the level of intelligence, the European Commission differentiates between narrow AI systems (current development status in 2019) and general AI systems that can perform human activities (European Commission, 2019a).

Machine Learning

- The Oxford Learner's Dictionaries defines Machine Learning (ML) as a type of AI where computers process big datasets to learn how to do tasks (Oxford Learner's Dictionaries, n.d.).
- According to Tegmark, it is the study of algorithms that improve through experience (Tegmark, 2018).

Deep Learning

- Cambridge Dictionary defines Deep Learning (DL) as "a type of AI that uses algorithms [...] based on the way human brain operates" (Cambridge Dictionary, n.d.).

Artificial General Intelligence

- According to Max Tegmark (2018), the AGI can accomplish virtually any goal, which includes intelligent learning; compared to the narrow intelligence of a chess program.

2.2   Understanding of the Technology

It is natural for humans to rate the difficulty of tasks compared to how hard it is to perform them. This can create a misleading picture of how hard it is for computers. For people, it is harder to multiply 12,345 by 54,321 than recognizing a friend in a picture. For computers, it is the other way around. The fact that low-level sensorimotor tasks are easy despite these requiring much computational resources is known as Moravec's paradox (Tegmark, 2018). The smallest known memory device in the wild is the genome of the bacterium *Candidatus Carsonella* ruddii, which stores about 40 kilobytes. Human DNA stores about 1.6 gigabytes (Tegmark, 2018). Compared with numbers of currently used machine memories, it shows that good computers can beat any biological system memory.

The memory of human brains works differently compared to computer memory: The way it is built and used as well. To retrieve information in a computer memory it is necessarily required to know where it is stored on the hard disk. To retrieve information from the human brain it is required to specify something about what is stored. Each group of bits in the computer memory has a numerical address to retrieve the information. It works like sending somebody to the bookshelf on the right side, to take the third book from the second shelf, and to read what is written on page 123. The human brain works more similarly compared to how people are using search engines. A request with a term or phrase is sent to the brain and something pops up afterward. Such memory systems are called auto-associative, as these recall by association instead of an address (Tegmark, 2018). Many AI researchers are working to figure out how to implement functions. For example, the goal of a machine translation is to implement a function inputting bits representing text in one language and outputting bits representing the same text in another language. Therefore, functions are implemented with NAND (not and) gates. This function inputs two bits and outputs one (Tegmark, 2018). This book will not go into deeper technical details of function and module building as this will not answer the research questions.

Decreasing hardware costs and further developments like parallel processing are increasing computer speed. The best parallel computer is the quantum computer. It is tough, not clear if it will be possible to build a commercially competitive quantum computer within the next decades. New algorithms have been developed to dramatically speed up specific calculations like cracking cryptosystems and training neural networks. A quantum computer can also simulate the quantum-mechanical system like atoms, molecules, or new materials (Tegmark, 2018). Quantum computing can create radical changes in the future. How does a computer learn? A pocket calculator can win an arithmetic contest, but it is not learning or improving speed. Every programed function computes in the same way. The ability to learn is the most interesting aspect of general intelligence. When the first pocket calculators and chess programs have been developed, humans created artificial intelligence. To learn, it must rearrange itself to get better and better at computing the function by obeying the laws of physics (Tegmark, 2018). John Hopfield showed that the network of interconnected neurons can learn in an analogous way: if certain states repeatedly occur, the machine will gradually learn these states and return to them from any nearby state (Sulehria & Zhang, 2007).

Neural networks transform biological and artificial intelligence and started dominating the AI subfield of machine learning (ML). A neural network is a group of interconnected neurons that can influence each other's behavior. The human brain contains as many neurons as there are stars in the galaxy (Tegmark, 2018). Each neuron is connected to thousands of others via junctions called synapses. A popular model for an artificial neural network represents the state of each neuron by a single number and the strength of each synapse by a single number. In this model, each neuron updates its status at regular time steps by averaging the input from all connected neurons and applying an activation function to compute its next state (Tegmark, 2018). John Hopfield demonstrated that Hebbian learning (deterministic rule combined with law of physics, which updates the synapses over time) allowed an artificial neural network to store a lot of complex information by simply being exposed to them repeatedly, which is generally called training (Sulehria & Zhang, 2007).

2.3   Current and Near Future State

Deep Reinforcement Learning
The company DeepMind, acquired by Google in 2014, used a simple and powerful idea called deep reinforcement learning, which is a classic machine learning technique inspired by behaviorist psychology, where a positive reward increases the tendency to do something again. DeepMind used this technique for playing the computer game Breakout. Like an animal getting a snack from the owner for doing tricks, DeepMind's AI learned to move the paddle to catch the ball, because this increases the chance to get more points soon. DeepMind combined this idea with Deep Learning (DL). The potential of this technique is not limited to virtual games, it works with robots and other machines as well. For example, a robot can teach itself to walk or swim without any help from human programmers with deep reinforcement learning. Everything what is needed is a system that gives points for making progress. To be faster and to reduce the risk of damaging themselves the first stage of their learning process may happen in virtual reality (Tegmark, 2018).

Deep Learning
A milestone was when the DeepMind's AI system AlphaGo won in the game Go against Lee Sedol in 2016. Within a year after defeating Sedol, a further improved AlphaGo system had won against all 20 top players in the world without losing. The used main idea by AlphaGo was to combine the power of deep learning with the logical power of GOFAI (Good Old-Fashioned AI) from before the deep learning revolution. DeepMind used a big database of Go positions from human play and games where AlphaGo had played a clone of itself. Additionally, they trained a deep neural network to predict from each position the probability that the

white color will win. A separate network got trained to predict likely next moves. Then DeepMind combined the networks and used a GOFAI method to identify the next move which will lead to the strongest position on the board (Tegmark, 2018). Deep learning is an AI approach for neural networks and was introduced by McCullough and Pitts at the University of Chicago in 1944. Neural nets are part of machine learning, where a computer learns to perform tasks by analyzing training data that has been hand-labelled by humans beforehand. An object recognition algorithm uses thousands of labelled images, e.g., cats to find visual patterns that match their labels (Hardesty, 2017).

Natural Language

Another area where AI makes immense progress is language. In 2016, the Google Brain team updated successfully the Google Translate service with deep recurrent neural networks to improve the older GOFAI system. Recent progress in deep learning for speech-to-text and text-to-speech conversation made it possible to speak on a smartphone in one language and listen to the translated result. Natural language processing is one of the most advanced fields of AI and further success will have a big social impact, as language is central for human beings. Deep learning systems are good translators but striving the (current) challenge of being trained by massive data sets, discovering patterns and relations, but without ever relating the words to something in the real world. For example, the system can understand that "king" and "queen" is like "husband" and "wife," but it does not know what it means to be male or female yet (Tegmark, 2018).

2.4 Challenges

It is controversial how the AI topic will impact humans and social life. Based on technological progress, it is obvious that even in the near-term AI might have an immense impact on how humans see themselves, long before AGI will play a central role. Human intelligence combined with AI can have the potential to make life even better: major improvements in science and technology, reductions of accidents, disease, war, injustice, etc. Scenarios like those in the movie Terminator are very unrealistic (Tegmark, 2018). The threats are not attacking machines, but it shifts distraction of the people from the real risks and opportunities of AI, which is defined as a real challenge.

There is potential that the offered benefits by AI may not be fully discovered by society. This can be a growing risk and cause unintended results when a good invention will turn into a threat. Another risk is the massive usage or misusage of AI technology. Increasing speed can influence the job market and distributions of costs and benefits in an extremely quick-growing way. As AI systems will generate new abilities and skills, it will require radical ideas (e.g., universal basic income) to adapt to the changes (Floridi et al., 2018).

To move to a successful AI era, some questions will need to have an answer (Tegmark, 2018):

- How to make AI systems more robust and prevent them from crashing and getting hacked?
- How to adapt the legal system to the fast-developing AI systems to be fair and efficient?
- How can AI systems (e.g., weapons) be smarter and prevent killing or destroying humanity?
- How can humans drive automation with AI systems without increasing unemployment or reducing income?

AI algorithms must be robust against manipulation so that when, for example, an image recognition system for weapons scans at airport luggage control stations can differentiate between a weapon, several neutral objects combined looking like a weapon, or a weapon neutralized by another close object. Robustness is the main criterion for IT security applications, but it is not a frequently used criterion in ML research papers (Bostrom & Yudkowsky, 2014).

Information technology has already had an important impact on many sectors of human life. The more humans rely more on technology, the more important it is to develop robust and trustworthy systems. Throughout human history, people have adopted a "learning from problems" approach. The discovery of fire led to the invention of fire extinguishers and fire alarms. The innovation of automobiles led to the development of seat belts and air bags. The focus on even more powerful technologies can reach a point where a single mistake can lead to a threat to the humanity like a global nuclear war or a bioengineered pandemic. As technology grows more powerful, humanity should rely less on trial-and-error approaches and move further on to more safety engineering. Therefore, humans need to become more proactive than reactive by investing in safety research to prevent accidents from happening even once. AI safety research defines four main areas to focus on: verification, validation, security, and control (Tegmark, 2018).

2.5   Opportunities and Risks

The new AI domains, where AI will have an immense future impact and the definition and implementation of AI principles and guidelines will play a key role in the future of humankind, are listed below.

AI for Space Exploration
Computer technology made it possible that humans can fly to the moon and send unmanned spacecraft to explore other planets. In the future, AI will help to discover and explore new solar systems and galaxies. Therefore, the software and the AI system must be without any errors and free from any kind of manipulation (verification), ensuring that the software fulfils exactly the expected requirements (Tegmark, 2018). End of May 2020, the cooperation between SpaceX and NASA made it possible to send the Dragon spaceship with two astronauts to the International Space Station (ISS) as a test run for future commercial journeys through space.

AI for Finance
Information technology transformed and influenced the finance system by allowing resources to be efficiently managed and relocated all over the world. An AI focus is to offer future profit opportunities from financial trading as well as the development of new currency systems. Already today, most of the stock market buying and selling decisions are done automatically by computer systems (Tegmark, 2018).

AI for Manufacturing

Computer technology changed the way of producing and manufacturing food and products. Controlled robots enhance productivity, precision, and speed. 3D printers can create prototypes or final products at the place where they are. AI systems can produce in time and may reduce food or energy waste. Industrial robots are assembling cars and airplanes and making changes to their supply chain (Tegmark, 2018).

AI for Transportation

Information technology and AI systems can optimize public transport and traffic control. In the USA, motor vehicle accidents killed about 35,000 people in a year, which is seven times more compared to all industrial accidents (CDC, n.d.). An unsolved ethical aspect of autonomous driving is how the machine will take a decision in emergencies. If the situation results in damaging property or killing people, on what facts will the system base its decision? Will the vehicle protect the driver or kill a human crossing the street? Besides verification and validation to prevent accidents, the system may also need control functionality (human-in-the-loop system) like a human operator for monitoring and changing the behavior if necessary (Tegmark, 2018).

AI for Energy

Intelligent systems control power generation and distribution with advanced algorithms for balancing production and power consumption worldwide. AI can keep power plants operating safely and efficiently. Smart grids can share and optimize energy in a more efficient way and will create new ways of application. Buildings can change to intelligent rooftop solar and home-battery systems (Tegmark, 2018). The technology company Microsoft developed with their "AI for Good" program new research fields in Earth, Health, Accessibility, Humanitarian Action, and Cultural Heritage to gain key knowledge to improve the existing world and to lead future markets. Traditional (energy) companies will need to take actions to transform themselves into data-driven companies.

AI for Healthcare

Digitalization of medical records has already improved the decision-making for doctors. Online services allow a fast client and doctor conversation and real-time diagnoses of digital images. In 2016, Stanford university study results (Conger, 2016) show that AI could diagnose lung cancer using microscope images better than human pathologists. In 2015, a Dutch study (Litjens et al., 2015) showed that the identification of prostate cancer using resonance imaging (MRI) was as good as the human doctors. AI systems will continue to support or replace doctors in the future, as surgeons are already using intelligent robotic systems. In 2020, based on the worldwide Coronavirus situation, AI technology is used within mobile apps for contact tracing, which rises ethical, legal, and societal issues (Coeckelbergh, 2020).

AI for Communication

Internet and the emerging Internet of Things (IoT) are improving convenience, efficiency, and economic benefits from connecting and bringing online people and products (lamps, thermostats, freezers, machines, bio-chipped farm animals, etc.). The "connecting everything" leads to another challenge for computer scientists: to secure the hardware and software against hacks and malicious software (malware) like worms or script files that exploit bugs in the software system (Tegmark, 2018).

AI for Laws

To cooperate and regulate life, humans developed laws to facilitate cooperation. AI can generate robojudges, which can apply the same high legal standards without succumbing to human errors like bias (skin color, gender, sexual orientation, religion, nationality, etc.), fatigue, or a lack of the latest knowledge. Robojudges could ensure that the first time in history, humans become equal under the law (Tegmark, 2018). If it is possible to copy robojudges, cases can be handled in parallel rather than in series with waiting periods. Safety and security will be an important topic to exclude bugs and hacks. In 2016, a study argued that a prediction software used in the USA was biased against African Americans and contributed to unfair sentencing (Angwin et al., 2016). Another challenge is to continuously adapt the law to the fast development of technology.

Therefore, more technical scientists will be required in nontechnical areas like law schools and governments (Tegmark, 2018). China developed a "smart court" system driven by AI technology with its aim to increase efficiency and fairness. Judges work together with AI systems on every case and if they do not agree with the AI system's decision, they must hand in a special explanation to change the automated sentence system (Pleasance, 2022). In this case, the AI system leads the decision and humans can only interfere after the AI system's decision was taken.

Another ongoing challenge is the dispute between privacy versus freedom of information. This starts by using a smartphone, visiting websites, and sharing data (facial recognition, IoT, etc.). The technological development generates permanent tension between privacy advocates and controlled dictatorship systems. AI systems are increasing their skills to analyze brain data from fMRI scanners to figure out if a person says the truth or not (Bhutta et al., 2015). Implementing this technology in courtrooms can enable faster trials and fairer judgments if the systems are reliable. How would the court and the robojudges decide if the AI systems are able to generate realistic fake videos, which will influence the judge's results? Another aspect is if and how AI research should be restricted by law. Continuous development can lead to the point that machines are able to grant rights. Who takes responsibility for an accident caused by a self-driving car? Is it the owner, manufacturer, passenger, or the vehicle itself? Is it a moral conflict to grant machines rights or will this already lead to machine discrimination?

AI for Intelligent Weapons

Nuclear weapons already create a lot of tension between countries. Futuristic AI weapons can have the possibility to increase even more the tension and threats. Who is supervising the development and regulating the level of science and research? Will AI-based weapons end the wars forever or will they destroy the whole of humanity? Will machines fight against machines in the future or will AI-powered drones and other autonomous weapon systems (AWS) hunt humans? Is there an accountability gap between the use of AWS and who can be held accountable (Umbrello, 2021)? Will a machine be more rational in making decisions than a soldier? Future intelligent weapons have the potential to take humans out of the loop via algorithms that decide who to target and kill. Many current weapon systems still require a human to execute the final decision about the target. There are tendencies to remove this human in the loop to gain speed in military activities. Therefore, it needs an AI safety to guarantee a zero-percentage fault tolerance to prevent nightmare scenarios. In the past, humans in the loop like Vasili Arkhipov, who made the final decision not to fire a nuclear torpedo from a Russian submarine toward the USA in 1962, prevented another World War by following his gut instinct. It is not clear how an AI system would have worked in that situation; would it follow the command, or would the system be able to predict the outcome of the order (Tegmark, 2018)?

Based on the concern of over 3000 AI researchers, six presidents from the Association for the Advancement of Artificial Intelligence (AAAI) and leading technology companies like Google, Facebook, Microsoft, and Tesla, signed an open letter and listed autonomous weapons as the next revolution of war besides gunpower and nuclear weapons. They will be cheap to produce and based on the assumptions of the researchers, the coming AI arm race will not benefit for a better humanity (Tegmark, 2018). Many leading scientists are requesting an international treaty to limit and define ethical AI research, banned fields and how the ban would be enforced. Based on scientists discussions (Tegmark, 2018), the most beneficial groups who will benefit of an arm race are not the superpowers, instead, it will benefit small rouge states and non-state actors like terrorists. Other concerns are intelligent killer drones, which can be sent in big, unmanned groups to eliminate their target. It is not clear how to control those future weapons yet. One approach can be to make those weapons more ethical like, for example, not to kill civilian people (Tegmark, 2018). Other threats without building intelligent weapons are hacked nuclear power plants, electricity systems, communication systems, industrial robots, or financial systems. AI systems can be used by themselves to develop intelligent intrusion detection systems and cyberwar defense (Tegmark, 2018).

AI for Jobs and Economy
As AI will affect many consumers, it will also have an impact on the future job market. In 2013, the wealth of the bottom half of the world's population is the same as the one from the world's eight richest people

(Oxfam, 2017). In the past, the economy kept growing and the average income was increasing for the wealthiest (mostly to the top 1%), while the income for the poorest (around 90%) stagnated (Tegmark, 2018). Economists agree that inequality is rising. Erik Brynjolfsson and his MIT collaborator Andrew McAfee argue that digital technologies push these inequalities in three different ways (Baskin, 2018; Tegmark, 2018): First, old jobs will be replaced with new ones, which require more (technical) skills. Second, they claim that an ever-larger share of corporate income went to the ones who own the companies in the last 20 years. Digital product copies (books, music, etc.) are easier to produce and require less production people, which moves most of the revenue directly to the investors. This influenced, for example, the change of the "Big 3" in the USA: General Motors, Ford, and Chrysler were almost identical to Google, Apple, and Facebook in 2014 (Manyika, n.d.). Third, the digital economy often benefits superstars more than everyone else. For software, people are willing to pay less or nothing anymore. Therefore, in the digital marketplaces, there is only room for some big players left.

When it comes to carrier advice for kids, scientists, and researchers do not have a common assumption about which jobs will have an employment guarantee. Many of the future jobs are most likely not existing today. General common proposals are professions in which machines are not good in, but how will this look like in 10–30 years? Before deciding about job education, the following questions shall be taken into consideration as a forecast once AI systems have taken over some jobs (Frey & Osborne, 2017):

- Does the future job require social communication skills and interaction with people?
- Does the future job require creativity?
- Does the future job require to work in an unpredictable environment?

Jobs that require structured and predictive actions, will not last long before getting automated. Examples, therefore, are warehouse workers, cashiers, train operators, line cooks, truck drivers, etc. However, being in the "safe-areas" like, for example, professional writer, athlete, fashion designer, actor, or filmmaker does not guarantee any job in the future as well. Automation is not the only challenge, as many people will need to shift to other job professions, which will cause strong competition between humans all around the world, where only some superstars will succeed (Tegmark, 2018): How can governments influence future success? Is the current education model with one or two decades of school and university study, before starting to work, still working in the future? Or will it be better if people will work some years, go back to school or university, and continue working afterward or will every job include a permanent education routine (Bates, 2016).

If humans and AI keep automating job tasks for the future and automated jobs will be replaced with even better ones, how will this influence the employment rate? This effect happened since the first Industrial Revolution. Job pessimists think that the usage of AI will be different, and many people will not only become unemployed, but unemployable. The argue is that the free-market salaries are based on supply and demand and high intelligent automation will decrease the human salary below the living costs. Based on history, the hourly cost of human workforce always followed lower-income countries or led to new machines (Tegmark, 2018). Blue-collar jobs were replaced with white-collar jobs and in the future the human mind can be replaced by AI and machines. In 2007, the Scottish-American economist Gregory Clark explained future job prospects with the imagination of two horses speaking about the invention of the automobile (Tegmark, 2018). The horses were worried about the impact of technological unemployment for them. To shorten the story, the horses were waiting for replacement jobs, but it happened the other way around. In the USA, horses were slaughtered and their population decreased from about 26 million in 1915 to 3 million in 1960 (Kilby, 2007). There is no prediction yet, if there will be any connection from this story to the human future.

Jobs provide people with income and purpose. If the unemployment rate will increase, the system, like the government by today, will need to create income and purpose for the citizens without offering any jobs (Tegmark, 2018). Or will there be a shared responsibility between humans to take care of others? Considering war and inequality worldwide in the past, the humanity will therefore need a big impact change, therefore. In 2016, professor Moshe Vardi from the Rice University saw the AI-powered technology as a moral imperative to save lives and MIT professor Erik Brynjolfsson spoke about at least to save half of all people from getting worse (Tegmark, 2018).

One solution can be to offer a basic income for all. Countries like Finland, The Netherlands or Canada are experimenting. Based on a research study in Finland, the results show that the unemployed persons improved their self-perceived well-being, but overall, it did not show effects on employment during the first year (Kela, 2019). The income can be money, but physical goods like a free apartment to live in or free public transport as well. It is also not known whether humans will start doing automated jobs again based on a high unemployment rate, even if AI systems work more efficiently than humans (Tegmark, 2018). Therefore, it will need more research and human-centered decision makers for the future. The build of the future economy system should include everyone, not only the technological AI leaders or political forces. Another approach could be to give purpose to the people without offering jobs.

In 2012, a meta-study analyzed that unemployment tends to have negative long-term effects on well-being. The analyzed retirement setting was a mix of both negative and positive aspects (Luhmann et al., 2012). Research in the field of positive psychology has identified factors that improve a human's sense of well-being and purpose. The researchers found out that some jobs can provide many of them (Duckworth et al., 2005):

• Social network of friends and colleagues
• Healthy and virtuous lifestyle
• Respect, self-esteem, and self-efficacy
• Sense of being needed and making a difference
• Sense of meaning by being a part of something
• Serving something larger than oneself

Compensation of those factors can happen outside the workplace by doing sports, learning, being with families and friends, teams, community groups, schools, humanist organizations, political movements, etc. To generate a successful working low-employment society, those factors will need to be analyzed and implemented in detail. Therefore, the additional involvement of psychologists, sociologists, and educators besides scientists and economists will have a positive impact (Tegmark, 2018). Microsoft operates their own AI Business School that offers training in AI strategy, AI ready culture, Responsible AI, and ethical principles. Besides technology, Microsoft is working in broader social areas by shaping the future. In March 2020, Microsoft announced the job offer "Senior Researcher—Affective Computing, Human Understanding and Empathy Group" (Microsoft, 2020), which requires a PhD in Computer Science qualification. The same requirement is wanted for a "Senior Researcher—Human Computer Interaction." Are those job positions, which will shape the future, all in the technical software field or does this require other professions as well? Fifty people, who were working in the field of Microsoft news production (MSN), have been replaced by AI in May 2020 (Baker, 2020). A Microsoft spokesman argued that they evaluate their business on a regular basis and increased investment can led to redeployment and job elimination. One affected person said that it is demoralizing to think machines can replace humans, but it is happening (Baker, 2020).

Human-Level Intelligence

It is not clear yet, if and when the AI development progress will eventually stagnate based on the difficultness to solve obstacles or if AI researchers will succeed with their goal of developing human-level AGI (Tegmark, 2018). When would it be theoretically possible to generate human-level intelligence earliest? Based on an assumption of a scientist, it would still not be enough to create the raw computational power that is needed, even though AI researchers would know how to build human-level AGI in the best way by using existing computer hardware. It is also unclear, how many floating-point operations per second (FLOPS) are required to simulate the brain and human intelligence. Therefore, it may also be unclear yet how to provide enough electricity power for long-term operations. It is also unclear if the brain can be simulated with a neural network model or if it will require a model of individual molecules or subatomic particles, which would increase the number of FLOPS dramatically (Tegmark, 2018). As a step toward AGI scientists are developing more brain-like hardware like neuromorphic chips.

2.6   Goals

Humans are striving for a sense of life, self-fulfillment, or reaching personal goals, what would it be that pushes AI technology forward? Does AI (or AGI/superintelligence) need goals by itself for future development? If so, who will define those goals in first-hand? And which goals will an AGI, or superintelligence define? Will humans agree to those goals? The origin of goals can be found in physics, which discovered that the current known laws of physics can be mathematically reformulated in a similar way. The goal-orientation of a person can be described by the trial of optimization, which can be an attempt to improve health, happiness, etc. If the nature tries to optimize something, it will be limited to the laws of physics (Tegmark, 2018). The evolution of life has executed optimization based on a set of rules like drinking when thirsty, eating when hungry, or avoiding things that hurt. Those rules supported the goal of replication but could also fail by external influence (like a rat is eating delicious-tasting poisoned food). The human mind understands those rules as feelings, which influence the decision-making versus the meta goal of replication. Humans rebel against their genes and the replication goal, for example, by using contraceptives. In the beginning, the brain was evolved merely to help copying the human genes. Evolution created smarter brains than genes, which made it possible for humans to understand the replication goal of the genes and to make different decisions like not to produce and raise more children than manageable. Still those higher-developed rules can get abused with drug addiction for example. During evolution, the authority changed from genes to feelings. Therefore, human behavior is not optimized for survival anymore and does not have a single defined main goal (Tegmark, 2018).

Under consideration of the human evolution, should AI systems have goals? Therefore, it is necessary to specify the meaning of goal in detail. Can machines have a goal-oriented behavior? For example, does a dishwasher have a goal by itself? Machines have goals based on the design (Tegmark, 2018), which is done by humans like cleaning dishes for the dishwasher. Most of the today's-built machines do have a goal-oriented design but not a goal-oriented behavior. Further goal-oriented entities on the planet earth are bacteria, plants, animals, metals, etc. Another example is a mouse trap, which has the human designed goal to catch a mouse, but it does not understand how a mouse looks like, it has no feelings and no goal-oriented behavior like choosing different kinds of food based on the favorite taste of the mouse.

As machines are getting more and more intelligent, it gets more important to align the machine goals with the human ones. In the case of superintelligence, where the AGI system has the ability to strive for goals, the challenge will be that the system is faster and better at defining and reaching the goals (Tegmark, 2018). This means that the risk of AGI is not malignance but competence and capability to accomplish the goal in the fastest way. Therefore, the humans need to make the AI to learn, adopt and retain the human goals beforehand. Yudkowsky (2004), an AI safety pioneer, created therefore the term "friendly AI." To learn the human goals, an AI system must not understand what people are doing, it must instead understand why people are doing it. This is a difficult task for the computer that can cause a lot of misunderstandings or misinterpretations by today (Tegmark, 2018). How to teach an AGI that humans do not like to vomit during car rides or that humans do not eat silver metal? Therefore, a detailed model of the world, as humans are understanding it, needs to be created beforehand. Current AI research is trying to enable machines to infer goals from behavior.

If a self-improving superintelligence with the ability to learn and adopt the human goals will exist, will it retain the goals? With increasing intelligence, the targeted goal can be changed of the superintelligence by getting a new and different understanding of the world model (Tegmark, 2018). Another unclear point is who defines the human goals. Is this a committee of all nations or representatives of the strongest one? Will this lead to a worldwide dictatorship or balance the wealth globally?

2.7   Human Life and Behavior

Human life can be defined in three stages (Tegmark, 2018): biological evolution, cultural evolution, and technological evolution. The first biological stage is unable to redesign its hardware (made of atoms) or its software (made of bits) during its lifetime. It is defined by the DNA and can change only within an evolution over many generations (e.g., bacteria). The second cultural stage can redesign its software: humans are able to learn new complex skills like languages, sports, and can update their (personal) goals and understanding of the world. The current human state can be defined as stage 2.1, as we can do minor hardware upgrades like implanting artificial teeth, knees, or pacemakers, but no radical changes like e.g., getting ten times taller. The third technological stage, which is not existing now, can redesign the software and hardware instead of evolving over generations.

2.8   Chapter Summary

This chapter gives an AI overview by starting to list and compare different terms and definitions in Sect. 2.1. It shows, how important it is to clarify terms beforehand, as there are not always unique definitions and understanding by humans about AI. It highlights that an AI system does not represent a robot or physical machine. The software can be operated on different kinds of hardware systems or platforms but does not require to have any physical shape. Section 2.2 gives an overview of different AI technology aspects. It shows the differences in data computation and progressing between humans and computer systems. Section 2.3 describes several AI technologies and highlights the challenge of generated bias. If the AI algorithm gets trained with an already biased data set whether that can happen with consciousness or not, it will not generate a "better human."

The next Sect. 2.4 lists some AI challenges. A current main challenge is that humans still see AI systems as robots or machines and think in physical threats or opportunities. As AI systems can be only software modules, it is recommended to shift the broad society discussions from an attacking Terminator robot to a broader ethics and responsibility discussion. Every citizen should take responsibility and proactive actions to shape "Friendly AI" systems. Like every technology, also AI can get abused and used in a not general friendly way. Section 2.5 describes the opportunities and risks of the AI technology. As humans stopped being goal driven.

Section 2.6 highlights the scenario, where future AI systems might overtake the definition of goals for humans. To generate a broad awareness within the society to develop and implement AI systems in daily life, it needs acceptance, transparency, and understanding. Section 2.7 gives an overview of a humans-based classification on evolution.

References

Angwin, J., Larson, J., Mattu, S., & Kirchner, L. (2016). *Machine bias*. Accessed February 3, 2020, from https://www.propublica.org/article/machine-bias-risk-assessments-in-criminal-sentencing

Baker, G. (2020). Microsoft is cutting dozens of MSN news production workers and replacing them with artificial intelligence. *The Seattle Times*. Accessed June 9, 2020, from https://www.seattletimes.com/business/local-business/microsoft-is-cutting-dozens-of-msn-news-production-workers-and-replacing-them-with-artificial-intelligence

Baskin, K. (2018). *How to approach the second machine age*. MIT Management Sloan School. Accessed February 3, 2020, from https://mitsloan.mit.edu/ideas-made-to-matter/how-to-approach-second-machine-age

Bates, S. (2016). *Computers Gone Wild: Impact and implications of developments in artificial intelligence on society*. Accessed February 3, 2020, from http://futureoflife.org/2016/05/06/computers-gone-wild

Bhutta, M. R., Hong, M. J., Kim, Y.-H., & Hong, K.-S. (2015). Single-trial lie detection using a combined fNIRS-polygraph system. *Frontiers in Psychology, 6*, 709. https://doi.org/10.3389/fpsyg.2015.00709

Bostrom, N., & Yudkowsky, E. (2014). The ethics of artificial intelligence. In K. Frankish & W. Ramsey (Eds.), *The Cambridge handbook of artificial intelligence* (pp. 316–334). Cambridge University Press. https://doi.org/10.1017/CBO9781139046855.020

Cambridge Dictionary. (n.d.-a). *Consciousness*. Accessed January 31, 2020, from https://dictionary.cambridge.org/dictionary/english/consciousness

Cambridge Dictionary. (n.d.-b). *Deep learning*. Accessed March 16, 2020, from https://dictionary.cambridge.org/dictionary/english/deep-learning

CDC Centers for Disease Control and Prevention. (n.d.). *Motor vehicle injuries*. Accessed February 3, 2020, from https://www.cdc.gov/winnablebattles/report/motor.html

Coeckelbergh, M. (2020). *Corona app: Ethical, legal, and societal issues*. https://www.derstandard.at/story/2000117457461/corona-app-ethical-legal-and-societal-issues

Conger, K. (2016). *Computers trounce pathologists in predicting lung cancer type, severity*. Accessed February 3, 2020, from https://med.stanford.edu/news/all-news/2016/08/computers-trounce-pathologists-in-predicting-lung-cancer-severity.html

Duckworth, A. L., Steen, T. A., & Seligman, M. E. P. (2005). *Positive Psychology in Clinical Practice, 1,* 629–651. https://doi.org/10.1146/annurev. clinpsy.1.102803.144154

European Commission. (2019a). *A definition of AI: Main capabilities and disciplines.* Accessed March 31, 2020, from https://ec.europa.eu/newsroom/dae/document.cfm?doc_id=56341

European Commission. (2019b). *Ethics Guidelines for Trustworthy AI.* Accessed March 31, 2020, from https://ec.europa.eu/newsroom/dae/document. cfm?doc_id=60419

Floridi, L., Cowls, J., Beltrametti, M., Chatila, R., Chazerand, P., Dignum, V., Luetge, C., Madelin, R., Pagallo, U., Rossi, F., Schafer, B., Valcke, P., & Vayena, E. (2018). AI4People—An ethical framework for a good AI society: Opportunities. *Risks, Principles, and Recommendations, Minds & Machines, 28,* 689–707. https://doi.org/10.1007/s11023-018-9482-5

Frey, C. B., & Osborne, M. A. (2017). The future of employment: How susceptible are jobs to computerisation? *114,* 254–280. https://doi.org/10.1016/j. techfore.2016.08.019

Hardesty, L. (2017). *Explained: Neural networks.* Accessed June 8, 2020, from http://news.mit.edu/2017/explained-neural-networks-deep-learning-0414

Kela. (2019). *Preliminary results of the basic income experiment: Self-perceived wellbeing improved, during the first year no effects on employment.* Accessed January 28, 2020, from https://www.kela.fi/web/en/news-archive/-/asset_publisher/lN08GY2nIrZo/content/preliminary-results-of-the-basic-income-experiment-self-perceived-wellbeing-improved-during-the-first-year-no-effects-on-employment

Kilby, E. R. (2007). *The demographics of the U.S. equine population.* Accessed February 3, 2020, from http://www.humanesociety.org/sites/default/files/archive/assets/pdfs/hsp/soaiv_07_ch10.pdf

Litjens, G. J., Barentsz, J. O., Karssemeijer, N., & Huisman, H. J. (2015). Clinical evaluation of a computer-aided diagnosis system for determining cancer aggressiveness in prostate MRI. *European Radiology, 25*(11), 3187–3199. https://doi.org/10.1007/s00330-015-3743-y

Luhmann, M., Hofmann, W., Eid, M., & Lucas, R. E. (2012). Subjective well-being and adaptation to life events: A meta-analysis. *Journal of Personality and Social Psychology, 102*(3), 592–615. https://doi.org/10.1037/a0025948

Manyika, J. (n.d.). *Can technology and productivity save the day?* Accessed June 14, 2020, from http://futureoflife.org/data/PDF/james_manyika.pdf

Microsoft. (2020). *Senior Researcher – Affective Computing, Human Understanding and Empathy Group*. Accessed June 9, 2020, from https://careers.microsoft.com/professionals/us/en/job/708826/Senior-Researcher-Affective-Computing-Human-Understanding-and-Empathy-Group

Oxfam. (2017). *Just 8 men own same wealth as half the world*. Accessed February 3, 2020, from https://www.oxfam.org/en/press-releases/just-8-men-own-same-wealth-half-world

Pleasance, C. (2022). China uses AI to 'improve' courts - with computers 'correcting perceived human errors in a verdict' and JUDGES forced to submit a written explanation to the MACHINE if they disagree, MailOnline. Accessed August 8, 2022, from https://www.dailymail.co.uk/news/article-11010077/Chinese-courts-allow-AI-make-rulings-charge-people-carry-punishments.html

Ransbotham, S., Khodabandeh, S., Fehling, R., LaFountain, B., & Krion, D. (2019). *Winning with AI*. MIT Sloan Management Review and Boston Consulting Group. Accessed November 28, 2019, from https://sloanreview.mit.edu/ai2019

Rothenberger, L., Fabian, B., & Arunov, E. (2019). Relevance of ethical guidelines for artificial intelligence – A survey and evaluation. In *Proceedings of the 27 European Conference on Information Systems (ECIS)*, Stockholm & Uppsala, Sweden, June 8–14, 2019. ISBN: 978-1-7336325-0-8 Research-in-Progress Papers. https://aisel.aisnet.org/ecis2019_rip/26

Siau, K., & Wang, W. (2018). *Ethical and moral issues with AI - a case study on healthcare robots*. Emergent research forum (ERF). Accessed October 20, 2019, from https://www.researchgate.net/publication/325934375_Ethical_and_Moral_Issues_with_AI

Sulehria, H., & Zhang, Y. (2007). Hopfield neural networks – A survey. In *Proceedings of the 6th WSEAS Int. Conf. on Artificial Intelligence, Knowledge Engineering and Data Bases*, Corfu Island, Greece, February 16–19, 2007. Accessed December 27, 2019, from https://pdfs.semanticscholar.org/ae11/03b32c83c9489f974d605d974d4d98c2562d.pdf

Tegmark, M. (2018). *Life 3.0*. Penguin Random House.

Umbrello, S. (2021). Coupling levels of abstraction in understanding meaningful human control of autonomous weapons: A two-tiered approach. *Ethics and Information Technology, 23*, 455–464. https://doi.org/10.1007/s10676-021-09588-w

Yudkowsky, E. (2004). *Coherent extrapolated volition, MIRI*. Accessed February 3, 2020, from http://intelligence.org/files/CEV.pdf

3

Ethics and Moral

An AI system can be abused by somebody with a lack of moral by only fulfilling the owner's goal. What will happen if this ethical goal alignment is not achieved before the finished development of a superintelligence? Max Tegmark writes that postponing working on ethical standards after a goal-aligned superintelligence is build, would be disastrous (Tegmark, 2018). Already in ancient time, philosophers started to work on ethic principles like Aristotle, Plato, or Immanuel Kant. There was discord about ethical themes like sexuality or equality, but also widespread agreement about the emphasis on beauty, goodness, and truth. A rule like you should treat others the way you would like to be treated by others, occurs in many cultures and religions. It intends to evolve harmonious continuation of the human society. This can be compared with ethical rules around the world like the Confucian emphasis about honesty or the Ten Commandments. Western ethics are based on several attitudes and responsibility frameworks including teleology (good action, good consequences, important approaches: utilitarianism, antiquity, and hedonism) and deontology (values and principles, what is right, important approach: virtue ethics) (Rothenberger et al., 2019).

J. Baker-Brunnbauer, *Trustworthy Artificial Intelligence Implementation*, Business Guides on the Go, https://doi.org/10.1007/978-3-031-18275-4_3

One of the first who discussed guidelines for intelligent systems was Isaac Asimov with "Runaround" in 1942, which turned into the "Three Laws for Robotics" later (Asimov, 2004):

- A robot may not harm a human being.
- A robot must obey orders.
- A robot must protect its own existence.

Future AI systems will operate more integrated with humans and may have their own moral status like being an own moral entity or doing tasks by its own will (Bostrom & Yudkowsky, 2014). What will be the future human (ethical) understanding of putting the moral status of an AI system above the will of a human? Many ethical principles have social emotions like compassion and empathy in common. The parameters are reward and punishment for guidance. If a human does something bad and feels badly about it, the emotional punishment is generated by the brain. If a human disregards ethical principles, the society may punish through shaming by peers or sentence at the court. In the past, all survived human societies had ethical principles with a focus on survival. What kind of ethical ruleset does humanity need for the future? Scientists see no common ethical consensus in today's world, but there are basic principles with a broad agreement (Tegmark, 2018):

- Utilitarianism: Maximize positive experiences and minimize suffering.
- Diversity: Usage of diverse sets of positive experiences.
- Autonomy: Entities should have the freedom to pursue their own goals.
- Legacy: Compatibility with scenarios that most of the humans see as positive.

The implementation of those principles is a challenge (Tegmark, 2018). Based on the understanding that every entity, which includes, for example, an alligator, is conscious, has the freedom to learn, think and communicate how would a future AI system be programmed? Would this lead to that all humans and animals would be vegetarians? If the alligator will kill another animal, would this be illegal? Will it need rules to terminate digital life forms as well? How would those look like and how to guarantee that those rules will be executed? How to control a digital population ethically in the future? There are implementation challenges in the defined legacy principle as well (Tegmark, 2018). As today's humans would not like to define and create the environment by a mindset of people from the middle age, would an intelligent AGI system consider the needs and understandings of lower-developed humans? Therefore, it needs more discussion, awareness, and further research about how to apply ethical principles in the future, as the research and development of technical AI systems keep progressing.

Defining applied ethics for cases with very different contexts (e.g., fast recursive reproduction) needs completely different ethical principles for foundational normative truths and it is important not to take current familiar human principles as standards. Irving John Good defined the term "intelligence explosion" by describing that an intelligent AI system can understand its own code and recreates a more intelligent version of itself, which is able to do the same (Good, 1965). Recursive scenarios are already existing by humans using augmented intelligence with brain–computer interfaces colliding (Bostrom & Yudkowsky, 2014). Humans (first general intelligence) on earth, reshaped the world by influencing the planet (rivers, mountains, skyscrapers, framing, pollution, climate changes, etc.). A more powerful intelligence can or will have even more impact on earth or beyond.

In 2006, the concept of machine ethics that was proposed by Anderson and Anderson started discussions about ethical issues. The AI focus is mainly on financial investment and technology, but the discussions about ethics and morality are still in the beginning. Siau and Wang (2018) describe ethics as a complicated and complex concept with a focus on a single aspect. Relevant literature broadened the view on ethics and analyzed the relations to AI topics by following scientists and researchers, as it is not possible to study ethical topics of AI under all situations:

- Belmont (1979) defined a framework of three points: Respect for subject (the right to decide), beneficence (do not harm), and justice (fairly distribute).
- Mason (1986) defined PAPA issues: Privacy, accuracy, property, and accessibility.

- Bentham (1996) speaks about act utilitarianism (consequences of each action first and case by case action) and hedonistic utilitarianism (pleasure and pain are the only consequences).
- Wallach (2014) defined three ethical principles: Fairness, accountability, and transparency.
- Sinnot-Armstrong (2015) uses consequentialism (actions that cause benefits than harm).
- Hursthouse and Pettigrove (2016) defined virtue ethics (having ethical thoughts and characters).
- Alexander and Moore (2016) speak about deontological ethics (conforming the rules, laws, and ethical duties).

The current used AI technology refers to Narrow AI or Weak AI and the ethical issues involve human interaction. A different situation exists when future AI systems will have the ability for its own moral status. Therefore, the AI system should not be threatened as a machine, rather than an object that has similar or equal rights as humans (Siau & Wang, 2018).

To avoid misusage of AI technology, the value of an ethical approach of AI technologies needs a strong focus and a compliance with an adopted law. Floridi et al. (2018) define the transformation of an ethical approach to AI ethics as "dual advantage." This means that one point is that ethics support organizations to take advantage of the AI-enabled social values (new opportunities which are socially acceptable). The other point is that ethics makes it possible for organizations to prevent costly mistakes. This prevention is an advantage to eliminate actions that are socially unacceptable but legally approved. According to Floridi et al. (2018), the dual advantage of ethics can only work in a setting of public trust and clear responsibilities. People will accept AI systems only if their outcome is seen as meaningful with a low risk level. The success will depend on public engagement with AI technologies, openness about operation, and how easy it will be to understand the AI systems for humans.

3.1   How to Train Ethics?

One of the most important factors to consider for AI algorithm trainings are the human bias like the gender bias or race bias. As AI systems need plenty of data to train the accuracy, those data sets are chosen by humans first hand. Existing biases may be transferred to AI systems, where they develop themselves for the future. Therefore, it is important to train algorithms without any human biases (Siau & Wang, 2018). If AI systems will get their own sentience in the future, will they generate their own biases?

Three potential ways to educate ethics to AI systems are highlighted (Moor, 2006):

- Implicit ethical agents: Forcing the machines actions to prevent unethical outcome.
- Explicit ethical agents: Quote explicitly the allowed and forbidden actions.
- Full ethical agents: Machines have consciousness, free will and intentionality.

Besides those three approaches, it is still open how to ethically interact with an AI system that have consciousness, emotion, moral sense, and feelings (Moor, 2006). Is it ethical to shut down (kill) an AI system, if it replaces human jobs? Is it ethical to use military AI systems? This is also connected to human moral values and ethics. An AI system can be a moral agent by being autonomy (machines without direct control of another agent), being intentionality (acting in a moral harmful or beneficial way and the actions are calculated by intention) and responsibly (machines fulfil the social role which includes assumed responsibilities) (Moor, 2006). The aspect to have an ethical status is based on two points (Torrance, 2011):

- Ethical productivity: The ones who do or do not do their duties, e.g., murderers.
- Ethical receptivity: The ones who benefit or get harmed by the ethical producers.

A human who is steering a car and must decide if the car (e.g., without working breaks) will hit a group of older people or switch to another lane and hit two children is first hand an ethical producer. This is a difficult situation for a human, but how would an AI system handle this situation and who is responsible for the results of this decision? Another example is military robots, which are current ethical recipients of human command. Is it ethical to decide over those robots? Today's human ethical and moral understanding maybe not suitable or correct for future understandings, which are like in the past (considering slavery). This leads to the challenges that humans cannot solve all the identified ethical problems and that humans cannot identify all (current) ethical problems (Siau & Wang, 2018).

Research shows that people accept a decision process by an algorithm that is based on rational principles. Another approach showed that decision-making by machines (robots) does not convince humans immediately (Indurkhya, 2019). To make machine behavior (moral decision-making) acceptable to humans, it needs explanations of a transparent psychological decision-making process (Indurkhya, 2019; Bostrom & Yudkowsky, 2014). One psychological challenge for AI ethics implementation is that humans apply different moral values for themselves as to others. Researcher discovered in research work about autonomous vehicles, which may sacrifice their passengers that humans would prefer to buy an autonomous vehicle that prioritizes to protect its passengers (Bonnefon et al., 2017).

A perfect solution would be to develop AI systems that follow ideal ethical principles (Anderson & Anderson, 2007). This sounds easy in theory, but it is difficult to realize in a worldwide environment. If developers would program AI systems not to harm humans, the system would need to understand the meaning of harm first. This would need a global level of ethics which includes a reduced set of information between ethical standard makers and AI developers (Siau & Wang, 2018). A requirement for moral AI systems is that humans accept their decisions and find them reasonable. This requires integrating broad human common sense, ethics, and values for the development of AI and especially AGI systems. Humans started to implement robotic and electronic components to their bodies and brains to increase their abilities for further development or to improve life standards. Researchers and technicians are working on robots with human look-alike and using biological materials. If more cognitive functions are implemented in those devices, it is not clear when the person will change to an android and will lose their own moral consciousness (Indurkhya, 2019).

The Deep Learning AI model GPT-3 from OpenAI makes decisions based on 175 billion parameters. The model is the successor to the language model GPT-2 for generating words. The researchers have conducted an analysis of biases to better understand their model regarding fairness and bias. Their study shows that Internet-trained models have Internet-scale bias (Brown et al., 2020). The model tends to reflect stereotypes based on the training data besides it includes 175 billion parameters. For examples, academic and higher-paid jobs were associated to male persons, and Christianity was associated with Ignorant, Judgmental, and Execution. Islam has been described in terms of terrorism, fasting and Allah; Judaism with words like racist, Arab or smartest (Brown et al., 2020).

3.2   Influence on Product Development

It can be distinguished between meta-ethics (clarifying the basic concept like what is right and what is wrong), applied ethics (deal with a set of moral dilemmas from a specific domain), and moral theories (different ways of the notion of prohibitions, permissions, etc.) like deontology (analyze the intent behind the action), utilitarianism (judging the action or behavior), or virtue ethics (what makes life good or bad) (Pagallo, 2017). The differentiation depends on the kind of moral theory. First, an ethical code under a defined moral theory needs to be aligned and then the creator (developer, management, etc.) can choose to use deontic logic or "principlism" and a theory of prima facie duties or a divine command logic (Bringsjord & Taylor, 2014). For deontic logic, the aim is to formalize and implement ethics in terms of what is permissible, forbidden, or obligatory. In the case of "principlism," the focus is on autonomy, beneficence, and doing no harm. The divine command logic goal is the ethical control of AI behavior (Pagallo, 2017).

AI algorithms that should replace human social functions must consider the following criteria during their development: transparency, incorruptibility, responsibility, auditability, and usability (Bostrom & Yudkowsky, 2014). This needs to be considered and implemented, especially for machine learning and upscaling processes, beforehand. Non-AI systems do not know their purpose. For example, a toaster does not know that the device should only toast bread, as it is representing the manufacturers' decision (intended purpose). It will work to toast paper or clothes too if the human operator (customer) decides. It is a challenge to develop a domain-specific AI system that will operate safely in several (thousands of) domains or in new ones which are not discovered or considered by humans yet. A future AGI system, which inspects a power plant software for software bugs, will need an understanding of a human engineer with ethical concerns, that is, to differentiate from ethical engineering (Bostrom & Yudkowsky, 2014).

AI and AGI ethics differ fundamentally from ethical disciplines of noncognitive technologies by following criteria (Bostrom & Yudkowsky, 2014):

- The specific behavior of the A(G)I system will not be predictable.
- Safety verification of the A(G)I system will be more difficult as it requires verifying what the system is trying to do (instead of safety testing from defined behavior in defined operating contexts).
- Engineering must include ethics.

At the University of Stanford, parts of the studies are about losing control of AI systems. Similar risks have been revealed by Elon Musk, Stephen Hawking, and Bill Gates. AI systems gain knowledge and skills from their interaction with the environment to evolve more cognitive structures (Pagallo, 2017). The Japanese government has already created special test zones "Tokku" for robot and AI testing. The primary aim of those zones is to install safe test environments for scientists as well as for the society to test and interact together and it fits the virtue ethic moral theory (Pagallo, 2017). Testing scenarios in open environments show new ways to tackle legal challenges, discover new risks and threats, and prevent possible losses of control of AI systems.

3.3   Ethical and Moral Challenges

In the past, some people believed that moral status was influenced by bloodline or caste. Nowadays, scientists do not believe that factors like assisted delivery, family planning, and in vitro fertilization or gamete selection are influencing moral status. Even a human clone should have the same moral status as all other humans (Bostrom & Yudkowsky, 2014).

- The principle of Substrate Non-Discrimination describes that if two beings have the same conscious experience and the same functions and differentiate themselves only in the substrate of their implementation, then they have the same moral level (Bostrom & Yudkowsky, 2014).
- The principle of Ontogeny Non-Discrimination describes that if two beings have the same functionality and the same consciousness experience and differentiate themselves only in the way how they came into existence, then they have the same moral level (Bostrom & Yudkowsky, 2014).

Researchers write about the unemployment challenge and that AI makes decisions that humans cannot understand and control anymore or get replaced completely. It is possible that AI systems can get access to personal data without the awareness of humans. If the AI system will make the decision by using machine language, how will humans control that? An example is therefore the development of autonomous vehicles (Siau & Wang, 2018). Matthew Wall (technology specialist at BBC) says autonomous driving is not only about engineering and technology, but also about human psychology. The same is valid for moral decision-making: it is not about logic and rationality it is about psychologically acceptable explanations. Therefore, it is important to understand how humans argue and to identify acceptable decisions (Indurkhya, 2019). AI technology like Deep Fake video or audio is modifying and replacing faces and voices in videos. Such algorithms are a potential threat to humans as they can, for example, manipulate political speeches, fake confessions, or modify news coverage.

AI moral challenges are, for example, future autonomous driving or AI military robots that should kill a terrorist surrounded by innocent people (Indurkhya, 2019). The last example can be difficult to implement, as also human soldiers normally follow their leader's command. How to handle the ethics and moral implementation of military AI systems? Can it be possible to generate and implement a global ethical standard or will the local interest overrule? Considering the past, there are many examples where humans were blackmailed and turned against their leader and changed their moral values (Indurkhya, 2019). This often happened during natural disasters like floods, fires, earthquakes, or in wars. During the Napoleonic wars (1803–1815), the triage system was used to handle the high number of injured soldiers with limited medical resources first. It continued to develop in the First World War and is nowadays a base for modern standards in hospital emergency departments (Christ et al., 2010). The manual process of making triage decisions has been optimized into algorithms that less professional medical staff can make a decision quickly (Larsen et al., 1973). When such decision procedures are accepted by the medical authorities, healthcare workers are trained to follow those procedures and do not need to set focus on moral decision-making (Indurkhya, 2019). How can a doctor deny medical treatment to a seriously wounded soldier? And how would an AI system decide for COVID-19 patients?

Another concern of researchers (Siau & Wang, 2018; Bostrom & Yudkowsky, 2014; Pagallo, 2017; Floridi et al., 2018) is accountability. When an AI system will do a failure, who will be responsible for that? The software developer, the data owner, the manufacturer, or the end user? The opposite of the AI technology benefits is significant risks for users, developers, and governments based on difficulties to explain the algorithms (black box) and the influences on data privacy and data security. One important challenge is how to handle ethical and moral challenges in combination with AI (Siau & Wang, 2018). Other challenges are privacy and security. The current research and development of AI systems is depending on a big amount of data, including personal and private data. To secure the data, each action should be recorded to prevent privacy-related risks (Siau & Wang, 2018).

Algorithms based on historical data can predict within a style, but cannot develop a completely new style. It can be possible to predict that a new style will emerge, but it is not possible to predict its content (Indurkhya, 2019). The researcher trained their software with the data of the UK top 40 singles chart from the past 50 years to analyze the popularity factors of those songs (Ni et al., 2011). An AI system with this data might predict the next winner of a future Eurovision competition, but it cannot predict changes in aesthetics or tastes (abstract art or atonal music). As an outcome, it shows that a machine learning approach with data from the past, cannot learn moral understandings like that slavery is immortal (Indurkhya, 2019). Another challenge for future prediction algorithms is that they do not consider the outcome and impact of the prediction for the future (Indurkhya, 2019). This can be compared to a court situation (Starr, 2014): if the judge needs to decide over a one-year or two-year prison sentence, the judge needs not only to understand the risk, it requires also to consider the recidivism risk of the defendant and how another year in prison will influence this. Current publicly known risk prediction algorithms do not include this additional information for decision-making (Indurkhya, 2019). It is not publicly available, if and how the Chinese AI court system handles this aspect (Pleasance, 2022).

Current moral norms need to be reconsidered based on the rapid reproduction possibility of AI. A broad AI system may be duplicating its software within (milli)seconds. The copy would be identical to its original and would be able to start copying itself immediately. Besides hardware limitations, the AI system can grow exponentially. The current ethical norm about human reproduction includes the understanding of reproduction freedom. The outcome is that every couple can decide by themselves, if they want to have children and if so, how many they want to have (focusing on Western countries and not including biological reproduction limitations). Another norm is that the society intervene if their parents are not able or refuse to provide the basic needs for their children. Under consideration of AI systems those two norms are already colliding (Bostrom & Yudkowsky, 2014).

One of the main AI challenges is taking responsibility. Therefore, guidelines and principles should help to reflect, and merge different stakeholder opinions like from AI experts and other broader audiences. This can be used in all stages of design, development, and deployment for modern AI societies. The term "artificial" may raise negative emotions like fear for humans as seen in modern science fiction literature or movies (Rothenberger et al., 2019). Also, to generate a universal guideline for different cultures will be a challenge that needs to get solved as AI systems will operate in different global areas and cultural settings. Researchers speak about three areas of ethical and social concerns: safety and errors, law and ethics, and social impact (Rothenberger et al., 2019).

3.4   Guidelines and Principles

The German society TÜV (technical inspection and product certification services) urges legal guidelines for the usage of AI systems in critical safety environments (TÜV-Verband Thüringen, 2020). Based on their own survey (Ipsos survey, $n = 1000$), German consumers require more transparency and safety for AI applications. Eighty-five percent of the participants say that AI systems should be only available after testing and certification from an independent third party. Seventy-eight percent of the participants see the German government in the lead role to pass laws. Only 17% of the interviewees said that they will trust the AI manufacturer regarding safety. Based on the survey, 94% of all attendees heard the term "Artificial Intelligence" before, but only 34% could explain the technology or its characteristics. Regarding the question of how tolerant people would be if the AI system would make a mistake, 40% expect a 100% correctly working system, and 17% said that failures and mistakes are normal. Compared to autonomous driving, 84% of the interviewees answered that the AI system must work without any failures. Why is the percentage in the autonomous car scenario much higher? Do the people have a better understanding of the use case? Would only 40% of the people like a law or judging system without failures and mistakes or are other use case scenarios still not imaginable for people? In the published position paper for governments (VdTÜV, 2019), one of the six quoted points requests an ethical standard for trust building of AI systems. The German society TÜV recommends that manufacturers should admit those standards to generate transparency and trust. It is neither mentioned to which ethical standards they refer nor who will define those and how.

For researchers, AI will have an immense impact on the future of humanity (Floridi et al., 2018). What is unclear, is who, how, when, and where will experience a negative or positive change. To tackle those questions, four main opportunities are considered for a future AI society:

- Who humans can become (autonomous self-realization)?
- What humans can do (human agency)?
- What humans can achieve (individual and social capabilities)?
- How can humans interact with others (societal cohesion)?

AI systems offer big optimization potential in for instance transportation and logistics or even preventing diseases and to reinvent society radically. Therefore, humans need to understand not to generate a dependency on (autonomous) AI systems and maintain the ability to influence final decision-making. The risk for humanity is that AI systems may lead to unplanned changes by the intention to make automated routines for people's easier life (Floridi et al., 2018). Organizations have stated the values and principles that should be used for the development and deployment of AI systems and technology within societies. Principles should serve as an ethical foundation for discussions, guidelines, standards, regulations, and laws. Researchers are focusing on commonalities and important differences in principles based on the following manifestos (Floridi et al., 2018):

- The Asilomar AI Principles (Future of Life Institute, January 2017).
- The Montreal Declaration for Responsible AI (University of Montreal, November 2017).
- The General Principles (Ethically Aligned Design: A Vision for Prioritizing Human Well-being with Autonomous and Intelligent Systems, December 2017).

- The Ethical Principles (European Commission's European Group on Ethics in Science and New Technologies, March 2018).
- Five overarching principles for an AI code (UK House of Lords Artificial Intelligence Committee, April 2018).
- The Tenets of the Partnership on AI (multi-stakeholder organization, 2018).

All listed manifestos together generate a summary of 47 principles. Overall, there is an overlap and coherence between the different approaches which is like the four core principles used in bioethics (Floridi et al., 2018): beneficence, non-maleficence, autonomy, and justice. The four bioethical principles adapt to the new ethical AI challenges. Based on the research outcome (Floridi et al., 2018), the following five ethical principles cover the meaning of the six expert-driven manifestos:

- Beneficence covers the following understanding: do only good, well-being, dignity, sustaining the planet, prioritize the human, common good, benefit humanity and empower people.
- Non-maleficence covers following understanding: (personal) privacy, security, do not harm, access and control data, misuse, and overuse awareness.
- Autonomy covers the following understanding: power to decide, right to make decisions (decision-making power), value of human choice, and decide to delegate.
- Justice covers the following understanding: preserving solidarity, promoting prosperity, eliminating discrimination, global justice, risk of bias in datasets, applying to correct the past, creating benefits, and preventing future harm.
- The extra and non-bioethical principle explicability covers the following understanding: enabling the other principles via intelligibility (how does it work?) and accountability (who is responsible for the way it works?). This principle complements the four bioethical ones and overall ensures that the AI technology, the organizations, and people who are developing this, are responsible for an event of negative outcomes.

Following principles for the category "Ethics and Values" have been aligned during the Asilomar conference on beneficial AI (Future of Life, n.d.): Safety, Failure Transparency, Judicial Transparency, Responsibility, Value Alignment, Human Values, Personal Privacy, Liberty and Privacy, Shared Benefit, Shared Prosperity, Human Control, Non-subversion, and AI Arms Race.

The High-Level Expert Group on Artificial Intelligence (AIHLEG) prepared the Ethics Guidelines for Trustworthy AI as part of the AI strategy of the European Commission, and it is based on three components (European Commission, 2019): lawful, ethical, and robust. Based on fundamental rights, four ethical principles and their values must be respected for the development of AI systems:

- Respect for human autonomy
- Prevention of harm
- Fairness
- Explicability

Besides, more vulnerable groups like persons with disabilities, children, people with risk of exclusion, or situations with asymmetry power should get involved. Developers should adopt adequate measures to defuse risks that are difficult to predict, identify, or measure. Based on ethical principles and fundamental rights, Trustworthy AI can be realized by seven key requirements by considering technical and nontechnical methods (European Commission, 2019):

- Human agency and oversight: including fundamental rights, human agency, and human oversight.
- Technical robustness and safety: including resilience to attack and security, fallback plan and general safety, accuracy, reliability, and reproducibility.
- Privacy and data governance: including respect for privacy, quality and integrity of data, and access to data.
- Transparency: including traceability, explainability, and communication.

- Diversity, non-discrimination, and fairness: including the avoidance of unfair bias, accessibility and universal design, and stakeholder participation.
- Societal and environmental well-being: including sustainability and environmental friendliness, social impact, society, and democracy.
- Accountability: including auditability, minimization and reporting of negative impact, trade-offs, and redress.

The European nondiscrimination law differentiates between direct (illegal and less favorable behavior) and indirect (comparison between people, disadvantages, seemingly neutral provision, etc.) discrimination. The scope of the European nondiscrimination law is about ethnicity, gender, religion or belief, disability, age, or sexual orientation. Automated discrimination done by AI algorithms is more abstract and unintuitive, tangible, and difficult to discover compared to conventional forms as there might be a lack of access to evidence (Wachter et al., 2020). Therefore, a proposed solution is that technical and legal communities are working together to enable consistent assessment without interpretation of cases of automated discrimination (Wachter et al., 2020).

In 2019, the first international workshop on Digital Humanism was initiated in Vienna, Austria. It was organized by the Faculty of Informatics of TU Wien and 100 attendees with different backgrounds formed the Vienna Manifesto on Digital Humanism, which is a call to deliberate and to act on current and future development to shape technology with human values and needs. The manifesto (DIGHUM, 2019) includes ten principles. Researchers from the Montreal AI Ethics Institute are working on a framework to measure the environmental and social impact of AI (Gupta et al., 2020). Therefore, the four-pillar framework named SECure considers the carbon and social footprint of AI algorithms that can require an enormous amount of computational power.

The European Commission offers a Trustworthy AI Assessment List that can be adapted for each specific AI use case with the aim to achieve a general framework (European Commission, 2019). The European Commission published a White Paper on AI, aiming to become a global leader in innovation in the data economy and its applications (European

Commission, 2020). The document proposes to optimize research, to foster collaboration between the European member states, and to increase investments in AI development. Besides, it drafts a future European regulatory framework to mobilize resources to achieve an "ecosystem of excellence" along the whole value chain and the key elements of the framework will create an "ecosystem of trust" (European Commission, 2020). After the publication in February 2020, the paper undertook an open and public consultation process, where all European citizens, member states, and relevant stakeholders can participate by giving feedback. The European Union Agency for Fundamental Rights (FRA) has collected over 290 AI policy initiatives in EU Member States between 2016 and 2020 (FRA, 2020).

In June 2020, the German company Continental announced to develop a code of ethics for their internal development and usage of AI. Their Chief Technology Officer said that smart algorithms play an important role and he sees Continental, as a technology company, responsible to ensure internal ethical standards for development and processes. Further, AI decision-making must be always nondiscriminatory, transparent, and understandable (Continental, 2020). The new internal ethic guideline will be based on the EC Trustworthy AI guideline (European Commission, 2019).

Researcher generated a dynamic list of 20 action points in four categories (assess, develop, incentivize, and support) as a recommendation to policy makers for a good AI society (Floridi et al., 2018). AI systems should be designed to decrease inequality, respect human autonomy, and should increase benefits that are usable for all humans. A highlighted point is that AI systems are explicable to build trust and understand of the technology. Another defined factor is the requirement of a multistakeholder approach to ensure that AI systems will serve society needs. Therefore, developers, users, and rule-makers need to work integrated and together. Overall, different cultural frameworks need to be considered as they will drive the technology in different ways. Researchers defined 20 action points under consideration of a European cultural approach (including secure peoples' trust, serve the public interest and strengthen and share social responsibility) (Floridi et al., 2018).

Assessment

- Capacity of existing institutions (e.g., national civil court) to eliminate mistakes done by AI systems to evaluate liability.
- Assess the non-delegated functionalities of AI systems to ensure alignment with social values.
- Assess if current regulations are sufficiently defined to keep the pace with future technology development speed.

Development

- Development of a framework to increase transparency in the decision-making process.
- Development of legal procedures and improvement of the justice system (IT infrastructure) to include algorithmic investigations in court.
- Development of an audit system for AI systems to identify unintended implications like bias.
- Development of a mechanism to correct a complaint of an AI system to gain public trust. This may include an AI ombudsperson, guided complaint processes, and new insurance mechanisms.
- Development of metrics of trustworthiness for AI systems (trust comparison index) to increase public understanding.
- Development of a new supervisory council that is responsible for the public welfare through scientific screenings and supervision of AI systems (post-release monitoring).
- Development of an (European) observation group for AI systems to watch developments, provide AI literature sources and discussion forums.
- Development of contract templates and legal support for human–machine and AI system collaborations.

Incentivization

- Financially incentivize the development of AI systems that generate social benefits and are environmentally friendly.
- Financially incentivize research to advance AI for social well-being.

- Financially incentivize cross-sectoral and cross-disciplinary partnerships and developments between technology, legal studies, social standards, and ethics. Generate diverse and multi-stakeholder groups to consider different perspectives.
- Financially incentivize the inclusion of social standards, ethics, and legal standards in AI research projects.
- Financially incentivize the generation of law-deregulated testing zones for empirical testing and development of AI systems (cf. Tokku in Japan).
- Financially incentivize research about public understanding of AI systems to support the future generation of standards, policies, rules, and public opinion research methods.

Support

- It should support the development of codes of conduct for AI systems and datasets. The creation of an "ethical AI" certification program can generate public trust and people can demand certified AI systems.
- It should support the corporate management (C-level) to take responsibility for ethical implications (ethics committee, training, audits, corporate ethical review board, etc.).
- It should support the education system and the public regarding the AI impact on society, which includes school and university curricula, qualification and training programs, adaption of technology studies (AI, data engineer, software development etc.), and programs for the public.

Researcher generated an overview of guidelines for AI ethics (Rothenberger et al., 2019). Therefore, different types of organizations (industry, academia, research and development institution, government, and association) got analyzed for their guidelines and focus fields. Based on this overview, a catalogue of guidelines and definitions has been created and ranked via interviewed experts and an online survey (number one has highest priority):

1. Responsibility: The user and operator of an AI system are responsible for any action and consequences.
2. Protection of Data Privacy: Prevent unauthorized interruptions and the user needs to agree to the usage of the private data explicitly.
3. Transparency: An AI system must be transparent about being an AI, which needs to be standardized internationally.
4. Robustness: AI systems and algorithms need to be robust against internal and external manipulations.
5. Minimized Biases: Eliminate unfair and racist bias.
6. AI Purpose: The highest AI purpose must be to support and not replace human. Define a human–machine cooperation model.

The outcome of the research results is classified into four paradigm types (Rothenberger et al., 2019): engineering view, research view, innovation view, and compliance view. They say that engineering and research are not very enthusiastic for launching AI guidelines. In the interviewee's opinion, there are already enough scientific approaches existing. The researchers mention that the increasing discussions about ethics in Germany are caused by a translation error of the term AI into the German language because people are comparing machine intelligence to human intelligence (Rothenberger et al., 2019). Outcomes show that innovation and compliance strongly need a system of rules. Regarding the highest rated criteria "responsibility" research participants were asked about who will be responsible for the actions and consequences. AI manufacturer, developer/programmer, or the user? Some research participants (mostly engineers) said that from their perspective no ethics at all is necessary but mentioned to follow existing engineering guidelines. The research results could not answer whether a general ethical guideline or a specific guideline in particular application domains will be required. Based on the expert arguments that AI systems will be specific for a special domain, it would not need a general ethics guideline, but several detailed versions based on the application domains (Rothenberger et al., 2019).

3.5   Summary

This chapter focuses on ethics and moral based on an AI environment. It sums up that all past societies had ethical standards with the central goal to survive. Ethics itself goes back a long time in history. Already Aristotle was speaking about ethics, and it will be an even more central topic for the future of humanity. Will AGI consider the needs and understandings of the lower-developed humans? How will humans handle "intelligence explosion" and recursive scenarios, where AI or AGI will replicate itself constantly? Humans reshaped the planet earth to gain benefits, will this be done by an AI system as well?

This chapter lists an overview of different AI ethics frameworks and lists different viewpoints for the situation that AI systems will have their own moral status. Section 3.1 describes approaches about how to train AI systems in ethics. One major challenge is, if AI systems will get trained with already biased data sets. Overall, it is still unclear how to teach ethics to AI systems. Section 3.2 lists criteria for Product Development of AI systems and describes the importance of testing environments. Next Sect. 3.3 analyses some ethical and moral challenges. It shows that moral decision-making is not about logic and rationality, instead it is about psychologically acceptable explanations. Therefore, it requires a deep understanding of humans and their behavior.

Other challenges are accountability (who is responsible for a failure?), law adaption, quick reproduction of AI systems, and the not predictable social impact. Section 3.4 dives deeper into existing guidelines. A survey shows the importance of the requirement of transparency and safety for users and consumers. It describes the outcome of the summary of an overview of 47 values and principles that are based on several manifestos. The AIHLEG of the European Commission defines Trustworthy AI by seven key requirements and offers an AI assessment list.

References

Anderson, M., & Anderson, S. L. (2007). Machine ethics: Creating an ethical intelligent agent. *AI Magazine, 28*(4), 15. https://doi.org/10.1609/aimag.v28i4.2065

Asimov, I. (2004). *I, Robot (the robot series)*. Random House.

Bonnefon, J.-F., Shariff, A., & Rahwan, I. (2017). The social dilemma of autonomous vehicles. *Science, 352*(6293), 1573–1576. Accessed February 29, 2020, from https://science.sciencemag.org/content/352/6293/1573

Bostrom, N., & Yudkowsky, E. (2014). The ethics of artificial intelligence. In K. Frankish & W. Ramsey (Eds.), *The Cambridge handbook of artificial intelligence* (pp. 316–334). Cambridge University Press. https://doi.org/10.1017/CBO9781139046855.020

Bringsjord, S., & Taylor, J. (2014). The divine-command approach to robot ethics. In P. Lin, K. Abney, & G. A. Bekey (Eds.), *Robot ethics: The ethical and social implications of robotics* (pp. 85–108). The MIT Press.

Brown, T. B., Mann, B., Ryder, N., Subbiah, M., Kaplan, J., Dhariwal, P., Neelakantan, A., Shyam, P., Sastry, G., Askell, A., Agarwal, S., Herbert-Voss, A., Krueger, G., Henighan, T., Child, R., Ramesh, A., Ziegler, D. M., Wu, J., Winter, C., Hesse, C., Chen, M., Sigler, E., Litwin, M., Gray, S., Chess, B., Clark, J., Berner, C., McCandlish, S., Radford, A., Sutskever, I., & Amodei, D. (2020). *Language models are few-shot learners.* Accessed June 9, 2020, from https://arxiv.org/abs/2005.14165

Christ, M., Grossmann, F., Winter, D., Bingisser, R., & Platz, E. (2010). Modern triage in the emergency department. *Deutsches Ärzteblatt International, 107*(50), 892–898. https://doi.org/10.3238/arztebl.2010.0892

Continental. (2020). *Ethics regulations for Artificial Intelligence.* Accessed June 9, 2020, from https://www.continental.com/en/sustainability/news/news-2020/ai-code-of-ethics-224686

DIGHUM. (2019). *Vienna manifesto on digital humanism.* Accessed June 9, 2020, from https://www.informatik.tuwien.ac.at/dighum/index.php

European Commission. (2019). *Ethics guidelines for trustworthy AI.* Accessed March 31, 2020, from https://ec.europa.eu/newsroom/dae/document.cfm?doc_id=60419

European Commission. (2020). *White Paper on Artificial Intelligence: A European approach to excellence and trust.* Accessed February 19, 2020, from https://ec.europa.eu/info/publications/white-paper-artificial-intelligence-european-approach-excellence-and-trust_en

Floridi, L., Cowls, J., Beltrametti, M., Chatila, R., Chazerand, P., Dignum, V., Luetge, C., Madelin, R., Pagallo, U., Rossi, F., Schafer, B., Valcke, P., & Vayena, E. (2018). AI4People—An ethical framework for a Good AI society: Opportunities. *Risks, Principles, and Recommendations, Minds & Machines, 28,* 689–707. https://doi.org/10.1007/s11023-018-9482-5

FRA. (2020). *The European Union Agency for Fundamental Rights: AI Policy Initiatives (2016–2020).* Accessed July 4, 2020, from https://fra.europa.eu/en/project/2018/artificial-intelligence-big-data-and-fundamental-rights/ai-policy-initiatives

Future of Life. (n.d.). AI principles. Accessed January 31, 2020, from https://futureoflife.org/ai-principles

Good, I. J. (1965). Speculations concerning the first ultraintelligent machine. In F. L. Alt & M. Rubinoff (Eds.), *Advances in computers* (Vol. 6, pp. 31–88). Academic. https://doi.org/10.1016/S0065-2458(08)60418-0

Gupta, A., Lanteigne C., & Kingsley, S. (2020). *SECure: A social and environmental certificate for AI systems*. Accessed July 21, 2020, from https://arxiv.org/pdf/2006.06217.pdf

Indurkhya, B. (2019). Is morality the last frontier for machines? *New Ideas in Psychology, 54*, 107–111. ISSN: 0732-118X. https://doi.org/10.1016/j.newideapsych.2018.12.001

Larsen, K. T., Jr., Vickery, D. M., Collis, P. B., & Folland, E. D. (1973). Triage: A logical algorithmic alternative to a non - system. *Journal of the American College of Emergency Physicians, 2*(3), 183–187. https://doi.org/10.1016/S0361-1124(73)80049-8

Moor, J. H. (2006). The nature, importance, and difficulty of machine ethics. *IEEE Intelligent Systems, 21*(4). https://doi.org/10.1109/MIS.2006.80

Ni, Y., Santos-Rodriguez, R., McVicar, M., & Bie, T. D. (2011). *Hit song science once again a science?* Fourth International Workshop on Machine Learning and Music: Learning from Musical Structure, Sierra Nevada.

Pagallo, U. (2017). When morals ain't enough: Robots, ethics, and the rules of the law. *Minds & Machines, 27*, 625–638. https://doi.org/10.1007/s11023-017-9418-5

Pleasance, C. (2022). China uses AI to 'improve' courts – with computers 'correcting perceived human errors in a verdict' and JUDGES forced to submit a written explanation to the MACHINE if they disagree, MailOnline. Accessed August 8, 2022, from https://www.dailymail.co.uk/news/article-11010077/Chinese-courts-allow-AI-make-rulings-charge-people-carry-punishments.html

Rothenberger, L., Fabian, B., & Arunov, E. (2019). Relevance of ethical guidelines for artificial intelligence – A survey and evaluation. In *Proceedings of the 27 European Conference on Information Systems (ECIS)*, Stockholm & Uppsala, June 8–14, 2019. ISBN: 978-1-7336325-0-8 Research-in-Progress Papers. https://aisel.aisnet.org/ecis2019_rip/26

Siau, K., & Wang, W. (2018). *Ethical and moral issues with AI - A case study on healthcare robots*. Emergent research forum (ERF). Accessed October 20, 2019, from https://www.researchgate.net/publication/325934375_Ethical_and_Moral_Issues_with_AI

Starr, S. B. (2014). Evidence-based sentencing and the scientific rationalization of discrimination. *Stanford Law Review, 66*, 803.

Tegmark, M. (2018). *Life 3.0*. Penguin Random House UK.

Torrance, S. (2011). Machine ethics and the idea of a more-than-human moral world. In M. Anderson & S. Anderson (Eds.), *Machine ethics* (pp. 115–137). Cambridge University Press. https://doi.org/10.1017/CBO9780511978036.011

TÜV-Verband Thüringen. (2020). *Verbraucher wollen Sicherheit und Transparenz bei Künstlicher Intelligenz*. Accessed March 31, 2020, from https://www.tuev-thueringen.de/unternehmen/presse/texte/artikel/tuev-verband-verbraucher-wollen-sicherheit-und-transparenz-bei-kuenstlicher-intelligenz

VdTÜV. (2019). *Vertrauen in KI-basierte Systeme schaffen*. Accessed March 31, 2020, from https://www.vdtuev.de/themen/politische_positionen_vdtuev/kuenstliche-intelligenz

Wachter, S., Mittelstadt, B., & Russell, C. (2020). *Why fairness cannot be automated: Bridging the gap between EU non-discrimination law and AI*. https://doi.org/10.2139/ssrn.3547922 and webinar Accessed June 10, 2020.

4

Business and Management Perspective

Many research studies about ethics and moral in an AI context are conducted in universities by academics. In parallel companies and especially start-ups are developing AI products and services at high-speed with a strong technical focus. This chapter lists an overview of the current AI state combined with ethical perspectives from companies and management. For the fast-growing AI world, it will be important to ensure that academic research transfer into companies and start-ups in the future.

4.1 Current State of AI Implementations

An international research report outcome from MIT Sloan Management and Boston Consulting Group is that 7 of 10 surveyed companies answered that they have had minimal or no impact from AI up to now (Ransbotham et al., 2019). Ninety percent of the interviewed companies that have made at least some AI investments, less than 2–5 received some business gains from AI within the last 3 years. This number increased to 3 out of 5, when companies with significant investments have been added. Overall, the result is that 40% of companies with significant

J. Baker-Brunnbauer, *Trustworthy Artificial Intelligence Implementation*, Business Guides on the Go, https://doi.org/10.1007/978-3-031-18275-4_4

investments do not get any AI benefits (Ransbotham et al., 2019). On one hand, AI offers many opportunities for companies, but on the other hand, it is an existential risk for existing companies.

The results of the analysis show that the difficulties in generating value from AI are organizational and not technologically based (Ransbotham et al., 2019). Companies which have a clear AI focus on data, technology, and tools are gaining less value than those who are actively aligned AI owners, process owners, and business owners, which means that the executives enable their companies to consume AI as much as they produce it. Also, companies with an IT focus on AI technology generate less value than those with a broad strategic focus. To generate business value with AI requires access to a big data pool, which meets quantity and quality requirements. In many cases, managers need to allocate raw AI input data across organizational silos (Ransbotham et al., 2019). The usage of AI can be revolutionary, but C-level must act in a strategic way by deciding what their company should not do (Ransbotham et al., 2019).

4.2   Business Strategy

In 2019, an international research report from MIT Sloan Management and Boston Consulting Group describes that an increasing number of leading managers see AI not only as an opportunity, but they also define it as a strategic risk (Ransbotham et al., 2019). The awareness of this risk increased from 37% in 2017 to 45% in 2019 and is even higher in China. Therefore, executives are concerned about competitors which are using new AI technology. They analyzed how organizations are integrating AI into their strategy (Ransbotham et al., 2019). The two main approaches are integrating AI with strategic digital initiatives and the AI focus on revenue generation instead of cost reduction. Companies, which are tightly integrating AI and digital initiatives are 12% points more able to gain revenue impact and 20% points more likely that they have gained revenue or cost impact. Researchers analyzed the organizational behaviors of companies, which gain value from their AI initiatives (Ransbotham et al., 2019). They found out that AI is not only about technology. Successful patterns are:

- Integration of the AI strategy within the overall business strategy.
- Pushing large, risky, and lots of efforts for AI projects with revenue growth focus instead of cost reduction initiatives.
- Alignment of AI production with AI consumption.
- Investments in AI talents, data, and process changes.

They identified two main ways how companies handle AI risks (Ransbotham et al., 2019):

- The ones who gained value from AI are more proactive and they make riskier and bigger investments based on a calculated strategy.
- The other one's focus is on AI but as one approach in a broader strategic portfolio combined with the focus on organizational ability to consume AI.

Research results show that companies that hire or rent AI talent gain more value from their AI initiatives than others, who are relying only on their own existing employees or on externals (Ransbotham et al., 2019; Baker-Brunnbauer, 2019). Currently, there is an immense focus on AI algorithm development, but to materially affect business it is important to have competent and willing AI consumers. Therefore, AI applications must have direct applicability. For an efficient AI solution, business users and developers need to collaborate to define the correct AI algorithms for the specific use case (Ransbotham et al., 2019). Successful leaders do not only embed AI applications in their business strategy, but they also try to use AI as an organizational initiative of a data and technology combination for their organizational behavior and the way of working (Ransbotham et al., 2019). Research results show that most AI success stories are based on improving existing business processes in sales, marketing, pricing, manufacturing, etc. (Ransbotham et al., 2019). They compare this success to improve the gas mileage of combustion engines in a future era of transportation (Ransbotham et al., 2019). They indicate for executives to consider how to reimagine and reinvent the process in an AI context, where they define the true AI potential to disrupt industries most. In the future, executives must also consider how AI may affect their human resource and talent strategy. Skill sets and profiles for future jobs will be different compared to today (Ransbotham et al., 2019). The central out-come of the research study is that AI initiatives must be core to the com-pany's business strategy to create significant value and scale (Ransbotham et al., 2019). If the current business strategy does not identify AI as a risk or as an opportunity it will most likely be a pivot.

4.3 Management Decisions

The following short statements will give an overview of opinions, actions, and decisions by management so far:

Ali Keshavarz (Vice President and Head of Analytics from Aetna) sees the potential of AI in the enhancing field of customer engagement by directly connecting with customers and increasing the user experience with, e.g., automated billing and claim processes (Ransbotham et al., 2019).

Shivaji Dasgupta (Managing Director of Deutsche Bank) speaks about future market changes based on AI, especially in highly regulated indus-tries. New competition comes from new and diverse industries, which are not based on old established rules like the online bank N26, Apple Pay, Apple Card, or Amazon Cash. Especially tech leading companies like Apple, Google or Amazon will have a benefit of AI systems by using their existing customers and their generated data (Ransbotham et al., 2019).

Hervé Coureil (Chief Digital Officer at Schneider Electric) uses a digi-tal framework based on four factors: digital offerings, customer experi-ence, operations, and cybersecurity (Ransbotham et al., 2019). AI is integrated into all of the four key themes and provides a structure in which AI initiatives are considered, prioritized, and designed. Therefore, one key initiative was to generate new capability owner roles, who are savvy people about AI (not AI specialists) with acumen in key business areas to help leaders to prioritize AI projects (Ransbotham et al., 2019).

JeongHee Kim (Leader of the AI Research Lab of the Hyundai Motor Group) says that productivity is important, but their first AI goal is to increase the customers' value. Therefore, Hyundai considers using AI to create better in-vehicle environments or to improve safety performance rather than to focus on cost reduction and efficiency optimization (Ransbotham et al., 2019).

Werner Boeing (Chief Information Officer at Roche Diagnostics) defined three sectors of work for their digital transformation agenda: value chain, customer experience, and products and digital enhancement of products through services. In all those sectors, Roche is using AI via predictive maintenance, customer engagement, or solving complex scientific problems (Ransbotham et al., 2019). Boeing says that if leaders think about AI like a balance sheet, then they are missing the point and propose to generate emotional attachments to the disruptive nature (Ransbotham et al., 2019).

Markus Noga (Senior Vice President of SAP Cloud Platform Business Service) highlights how important it is that the AI software should not work as a black box in making decisions and actions (Ransbotham et al., 2019).

Sebastian DiGrande (Strategy and Chief Customer Officer at Gap Inc.) describes the AI future as an existential threat. Therefore, companies need to change the way they operate, the way they use tools, the level of automation and impact. Companies need to be aware of the situation that the customers and the industry will move on without them anyway, but they can try to influence it as much as possible (Ransbotham et al., 2019).

4.4 Summary

To build the bridge from academic research to companies that develop and use AI technologies for their product, service, or customer experience, this chapter focuses on C-level perspective. Section 4.1 describes the current state of AI implementations. Therefore, an international management report shows that most of the companies that have invested in AI did not get any impact from it up to now. Another aspect is that difficulties in generating value from AI are organizational and not technologically based. Another literature research outcome shows that companies with a clear AI focus on data and technology are gaining less value than those who actively align AI owners, process owners, and business owners. Also, companies with an AI IT focus generate less value than those with a broad strategic focus.

(continued)

(continued)

Section 4.2 goes deeper into business strategy. It describes that an increasing number of managers see AI not only as an opportunity, but they also define it as a strategic risk, which has been increasing over the last years. The two important strategic approaches are integrating AI with strategic digital initiatives and the AI focus on revenue generation instead of cost reduction. For the future, management needs to focus on human resource and talent strategy to ensure the development of AI systems. Section 4.3 gives an overview of some C-level statements about their AI implementation experience and status.

References

Baker-Brunnbauer, J. (2019). *Business model innovation in a paradoxical area of conflict (executive summary).* https://doi.org/10.13140/RG.2.2.24272.66566

Ransbotham, S., Khodabandeh, S., Fehling, R., LaFountain, B., & Krion, D. (2019). *Winning with AI.* MIT Sloan Management Review and Boston Consulting Group. Accessed November 28, 2019, from https://sloanreview.mit.edu/ai2019

5

Empirical Study

The outcome of the literature research shows that definitions and AI guidelines are differently understood between humans as well as within countries and cultures. Many papers describe AI technologies, social AI topics, and different approaches at academic research level. Nevertheless, there is no signed, committed, and implemented common European or global AI guideline now. Therefore, it depends on the countries themselves, if and how they push the AI technology. For the moment, companies of all maturity levels are developing AI systems based on their idea to generate business and to shape AI technology applications. Addressing ethical and moral topics related to AI is still in an early stage as it is not about "good or bad" or "right or wrong" (Siau & Wang, 2018). Moral and ethical issues regarding AI are critical and important for the society. Therefore, it needs an open discussion on different levels and a broad general understanding of the benefits and the risks. To build a bridge between academic research and AI business drivers, this empirical research analyses the understanding and awareness about the social impact of AI products and services from a management perspective.

© The Author(s), under exclusive license to Springer Nature Switzerland AG 2023
J. Baker-Brunnbauer, *Trustworthy Artificial Intelligence Implementation*, Business Guides on the Go, https://doi.org/10.1007/978-3-031-18275-4_5

5.1   Research Methodology

The focus of this book is to answer the question *"What kind of awareness does the management have about the social impact of their Artificial Intelligence (AI) product or service?"* by conducting expert interviews. Therefore, the focus is on the geographic central European area without any consideration of international cultural influence. The aim of this research is to further investigate the management understanding by answering the research question based on five sub-questions. First, to create a fundamental background and an understanding of the current research and literature state (Floridi et al., 2018; Pagallo, 2017; Indurkhya, 2019; Ransbotham et al., 2019; Siau & Wang, 2018; Bostrom & Yudkowsky, 2014; Tegmark, 2018; Rothenberger, Fabian & Arunov, 2019), research about moral and ethics in AI and their methodologies has been done. In the second step, the empirical research was done by qualitative expert interviews and has been analyzed via content analysis by following the approach of Mayring (2015) and Kuckartz (2018). Based on the research question and the outcome of the literature study, the following main categories have been defined for generating a structured interview guideline.

Motivation
The main reason why management started to develop an AI product or service.

Terms and Understanding
The used terms and understandings of "moral, ethics and artificial intelligence" within the management.

Prioritization
Describes how important and how deeply anchored are the three terms for the management and companies in their daily business.

Stakeholder Plus Profession
All internal and external people with their professions who are involved in the design and development of the AI product or service within the company.

Guideline and Manifesto
Collection of points, rules, or framesets that summarize or describes how ethics can be implemented in AI systems for shaping a good future society.

Certification
Willingness and understanding of AI system certification.

Ethical and Social Aspect for Societies
All ethical and social functionalities, activities, or changes in AI systems that will have a strong impact on shaping future societies.

Ethical and Social Aspects of Their AI System
All ethical and social functionalities, activities, or changes that influence the design and development of AI systems.

Interviewee Reflection
Learnings for the participant in the interview and additional input that is relevant and important to highlight from the management perspective.

Nine interview candidates have been invited to participate in the expert interviews. The goals of the expert interviews were:

- Analyze the current understanding and awareness about the social impact of AI systems from a management perspective.
- Analyze motivation, stakeholders and their professions, guidelines, and manifestos.

5.2 Definition of the Material

The used source materials are nine recorded expert interviews of executive managers (sampling unit, $n = 9$). The conducted interviews are based on predefined and pretested interview. The interview participants are mainly working as executives in different AI companies (Table 5.1). The mix of different business industries and countries generated diversity and collected different perspectives and insights.

Table 5.1 Interviewee constellation and participant's profile

Interviewee ID	Interviewee description	Company description	Country
I01	CEO and CTO for start-ups, master's degree in telecommunications, background in electronics, and communication	AI start-up focuses on AI and big data sector, data activations, and data marketplace platforms (data collection, data annotations, labelling, reselling of data)	Sweden
I02	CEO for two start-ups, master's degree in business administration, IT manager for 20 years in the banking and telco industry	AI start-up focuses on automatic web analytics and AI consulting, automated analytical process for website monitoring to generate suggestions, predictions, text analytics, natural language process, and recommendations for improvements	Denmark
I03	CDO and VP, master's degree in industrial engineering and Dipl.-Wirtsch.-Ing, founder of 3 companies	Healthcare company with more than one million registered users, digital platform for medical hardware products like blood pressure, blood oxygen, etc. using AI for pattern recognition in measurements. Working on an AI self-learning robot project	Germany
I04	CEO, around 20 employees	AI as a service product offering for different industries, using natural language processing, sensor data for predictive analytics, etc.	Finland

(continued)

Table 5.1 (continued)

Interviewee ID	Interviewee description	Company description	Country
I05	CEO and founder, a university degree in computer science	Deep Tech AI start-up, AI solution development and research, working with generative networks for anonymization and synthetization of large data sets for autonomous driving, reinforcement learning	Austria
I06	CTO and Co-founder, master's degree in science education and computer science	Scale-up, software as a service solution for e-commerce (online marketing and retail companies), using machine learning to predict optimal bids for the auction system from the Google Search Engine	Austria
I07	CTO, PhD in telematics, and master's degree in innovation management, responsible for big data infrastructure and business development	Research company, business, and projects between science and business, data-driven business consulting, text analysis, semantic map, open link data, industry 4.0 data analytics, domain agnostic learning	Austria
I08	Technical director, reasoning navigator, PhD in mathematics of physics, long-term strategy, and technology decisions	Digital health company (between start-up and established company), around 300 employees, app for patients to check their symptoms, information generation via AI engine	Germany
I09	Co-founder, studied business psychology, AI strategies, and data-driven business models	AI consulting for international enterprises, consulting in AI strategy and product innovation like, e.g., smart speakers or smart buildings	Germany

The selection process of the interview candidates was random by fulfilling the criterion that the potential candidate must be on an executive level of an AI company. Overall, the interviewee had to answer 19 questions and the interviews took place online between April and June 2020. The audio data of the interviews have been recorded and transcribed. The average duration of the recorded interviews is 40 min, resulting in around 100 pages of transcriptions. The interview was conducted by a guideline based on theory-driven categories to answer the central research question. The order of the questions stayed constant, and the interviews contained open questions. Based on the linguistic material, it is possible to generate different statements, like describing the approached topic, to find out something about the actors or the text's impact on the target group. The extended content analysis communication model is used to specify the direction (Mayring, 2015). Therefore, the material has been defined as analysis focus without considering nonverbal communication, gestures, mimics, laughers, or harrumphs. Through the interviews, the participants should be encouraged to reflect on their current understanding, awareness, and existing actions about ethics in AI for their company. The questions trigger thoughts about future development, social impacts, and how to implement AI technology for common good. Main and subcategories have been generated for the deductive coding process. Therefore, the categories based on the central research question and literature that have been used for the generation of the interview guideline, have been updated and extended.

References

Bostrom, N., & Yudkowsky, E. (2014). The ethics of artificial intelligence. In K. Frankish & W. Ramsey (Eds.), *The Cambridge handbook of artificial intelligence* (pp. 316–334). Cambridge University Press. https://doi.org/10.1017/CBO9781139046855.020

Floridi, L., Cowls, J., Beltrametti, M., Chatila, R., Chazerand, P., Dignum, V., Luetge, C., Madelin, R., Pagallo, U., Rossi, F., Schafer, B., Valcke, P., & Vayena, E. (2018). AI4People—An ethical framework for a good AI society: Opportunities. *Risks, Principles, and Recommendations, Minds & Machines, 28*, 689–707. https://doi.org/10.1007/s11023-018-9482-5

Indurkhya, B. (2019). Is morality the last frontier for machines? *New Ideas in Psychology*, *54*, 107–111. ISSN: 0732-118X. https://doi.org/10.1016/j.newideapsych.2018.12.001

Kuckartz, U. (2018). *Qualitative Inhaltsanalyse. Methoden, praxis, Computerunterstützung* (4th ed.). Beltz.

Mayring, P. (2015). *Qualitative Inhaltsanalyse: Grundlagen und Techniken* (12th ed.). Beltz.

Pagallo, U. (2017). When morals ain't enough: Robots, ethics, and the rules of the law. *Minds & Machines, 27*, 625–638. https://doi.org/10.1007/s11023-017-9418-5

Ransbotham, S., Khodabandeh, S., Fehling, R., LaFountain, B., & Krion, D. (2019). *Winning with AI*. MIT Sloan Management Review and Boston Consulting Group. Accessed November 28th, 2019, from https://sloanreview.mit.edu/ai2019

Rothenberger, L., Fabian, B., & Arunov, E. (2019). Relevance of ethical guidelines for artificial intelligence – A survey and evaluation. In *Proceedings of the 27th European Conference on Information Systems (ECIS)*, Stockholm & Uppsala, June 8–14, 2019. ISBN: 978-1-7336325-0-8 Research-in-Progress Papers. https://aisel.aisnet.org/ecis2019_rip/26

Siau, K., & Wang, W. (2018). *Ethical and moral issues with AI – A case study on healthcare robots. Missouri: Emergent research forum (ERF)*. Accessed October 20, 2019, from https://www.researchgate.net/publication/325934375_Ethical_and_Moral_Issues_with_AI

Tegmark, M. (2018). *Life 3.0*. Penguin Random House UK.

6

Findings

This chapter describes the outcome of the empirical research. Therefore, the results are structured and grouped to give answers to the defined sub-questions. The guidance was that the interview participants are speaking about their AI product or service and not about a general AI perspective.

6.1 Definitions

The main category "Terms and Understanding" and its subcategories "Moral, Ethics and Artificial Intelligence" are referring directly to the first sub-question (cf. Sect. 1.1) to answer the central research question. The findings of the interviewee's answers generate an insight into whether the managers are having the same understanding of the terms, as the outcome is influencing the creation of global and national principles and guidelines.

6.1.1   Moral

Interviewees describe the term moral as a guidance framework of principles that describes how humans are behaving in daily life based on how
they have been raised on their value system and culture. It describes a rule
set of acceptable behavior for others and defines the person itself by an
individual compass and loyalty. Moral and ethics are often used as syn-
onyms, but moral focuses on the operational and actionable aspects.

6.1.2   Ethics

Interviewees describe the term ethics as an umbrella or meta thing, where
morality is positioned under ethics. Ethics is a guidance framework of
principles that describe how to act and behave in life, and it helps to
develop moral. This ethical framework is society and culture specific and
defines the common way of actions and acceptable behavior within the
society. Ethics is related to the personal values and all humans should follow ethical guidelines, especially in the technology domain.

6.1.3   Artificial Intelligence

Interviewees describe the term artificial intelligence as a software system
that can perform tasks like object/image recognition, predictions, or text
analysis in a way that humans would do. AI makes machines intelligent
and autonomous to take over jobs from humans. Most AI approaches
learn from data and teach computers to think and make decisions on its
own. An interviewee describes AI by the status that humans do not
understand the status of the AI system anymore and that the term is
changing. AI is associated with machine learning, deep neural networks,
and supervised learning. AI is also used in the context of science fiction,
and it is an embedded system that is not tangible.

6.2   Prioritization

The main categories "Prioritization" and "Motivation" are referring directly to the second sub-question (cf. Sect. 1.1) to answer the central research question. The findings (open question and scoring) of the interviewee's answers generate an insight into whether the managers are setting focus on AI ethics and additionally describe their motivation about what drives them to develop an AI system. The interview participants have been asked how they prioritize ethics for their AI system. Firstly, they estimated a score between 1 (low) and 10 (high), where 77.77% of the interviewees rated the prioritization as high (>=8). Therefore, the calculated mean value is 8.22.

Secondly, the answers to the open questions: AI systems are scalable, and a wrong decision can have an immense impact in a very short time. Depending on the application field, like in medicine, mistakes, and unsuitable actions done by AI systems can be disastrous and require a careful implementation from day one. Ethical guidelines are installed within the company implicitly, but not explicitly defined. Ethics of AI and decision-making are covered by the company values. Ethics is a moral principle and therefore, the AI system should not harm. Younger employees in the company have an ethical approach focus and would not work for or within military projects or would leave the company. If the company does not act ethically, the team would not stay within the company, and it would also harm the brand. The highest priority, in the beginning, has the customer experience and ethics is increasing trust from your customer and employees that should not be lost. With AI systems it is possible to suggest the best marketing efforts and depending on the field of application, making wrong ethical decisions by AI is not very strong and the risk is low. Ethically AI solutions are important as they can collect data and build algorithms in a transparent way.

The motivation to start developing AI systems is based on reasons like curiosity about how AI will change humanity within the next 5–10 years, to build human-centered (serve and do not hurt humans) AI solutions, analyze, and interpret data with, for example, machine learning algorithms to predict a future state. Another motivation was customer complaints about online campaigns, direct customer inquiries or the creation of a better customer experience. For the medical company, the often-misdiagnosed rare diseases and to help patients to get directly the medical information was the motivation. Other reasons were to establish a top data science company, being frustrated about existing solutions and to create more intelligent ones. Another approach was to fight against fake news and misinformation by creating a transparent non-biased data marketplace.

6.3   Stakeholders

The main category "Stakeholder plus Profession" is referring directly to the third sub-question (cf. Sect. 1.1) to answer the central research question. The findings of the interviewee's answers generate an insight into who is involved during the AI system development process. Further insights are generated by the segmentation of internal and external stakeholders as well as the listing of their professions. The answers will demonstrate if there is an existing diversity or if the AI systems are mainly developed by technical IT people.

Nearly, all team members have psychological backgrounds as they are using a human-centered system approach. Their direct external stakeholder is the customers' IT department which often thinks less about ethics and AI. Internal stakeholders are described as AI scientists with a technical background in computer science, engineering, data scientists, software engineering, mathematics, bioinformatics, statistics, and physics. Other mentioned professions are medical doctors, UX designers, and researchers with a psychology background, project and company managers, and legal advisers. Business developers and customer meetings with sales and marketing are generating customer and market feedback for the AI system development. People diversity is generated by customer meetings and on the internal management level. People with social studies (psychology, philosophy) have been consulted during the company's founding period, but not anymore. Other psychologists are working in the team for human–machine interaction projects, other projects, and in the field of design and marketing. The arguments why companies are not hiring more people with social studies backgrounds are: company size, philosophy is not matching for the current AI development, but it can become a topic within this decade, it is not important enough for the business now or it is not considered by the management. One interviewee describes the company as open and that they are open minded to try new professions if it will fit within the business. External experts and external companies are as well involved in the AI system development. Young scientists have been chosen by the management, as they have a better understanding of the technology. Stakeholders have a research-oriented background and depending on the industry, the people's diversity can strongly change.

6.4   Guidelines and Principles

The main categories "Guidelines and Manifesto" and "Certification" are referring directly to the fourth sub-question (cf. Sect. 1.1) to answer the central research question. The findings of the interviewee's answers generate insights into whether the ethical guidelines and principles are known and used within their company. Therefore, it is segmented into advantages and disadvantages from the management perspective. Interviewees were also asked about who should take the lead in creating ethical AI guidelines and manifestos. The second part shows the findings about the willingness and thoughts of an ethical AI product or service certification.

6.4.1   Advantages of Ethical AI Guidelines

One company established its own manifesto for human-centered solutions that describes always to start with the person to understand the problem instead of choosing a technology approach and focusing on what is beneficial to humans. The goal of the manifesto is not to maximize happiness, it is about reducing pain for large parts of the society. Some companies have a company value document that has an internal universal character. An ethical AI guideline is not used within the company, as their product is far away from violating human rights or to harm people. One interviewee pointed out the awareness of generating readable and testable software code that is easy to understand in the future. During the interview, participants realized the internal need to create an ethical AI guideline for the company. Several interviewees mentioned the importance of a practical transfer of (AI) guidelines that can support the company and the employees to live up to these principles.

6.4.2   Doubts About Ethical AI Guidelines

Many interviewees answered that they do not have any explicit AI guidelines, but some interview participants spoke about an implicit way. Some managers said that they also currently see no need, the topic is new and less important, and they also will not create an AI guideline within the near future.

6.4.3   Guideline Proposals

For the analysis of one interview question, a different approach has been applied: Rothenberger et al. (2019) surveyed different ethical guidelines for their industry relevance. Therefore, they analyzed literature and conducted qualitative expert interviews. Their research overview generated a catalogue of six guideline proposals, which were ranked by seven experts during the interviews (Rothenberger et al., 2019). One question for the nine interview participants of this research work was to prioritize and to order the six ethical guideline proposals. For comparison, the arithmetic mean has been calculated (Table 6.1).

6.4.4   Ethical AI Guideline Creators

The creators and main actors should be parts of the society from every country and nation or an independent organization like Open AI Inc. This requires a general understanding and interest in AI technology that

Table 6.1 Comparison between the rank order of the guideline proposals

Guideline proposals	Arithmetic mean (Rothenberger et al., 2019)	Rank order (Rothenberger et al., 2019)	Arithmetic mean	Rank order
Responsibility	4.71	1	4.11	1
Protection of data privacy	4.43	2	3.89	2
AI should have a purpose	4.43	5	3.89	2
Transparency	4.14	2	3.67	3
Bias should be minimized	3.67	4	3.22	4
Robustness	3.29	3	2.22	5

gets updated, based on the fast-developing AI technologies, regularly. This also includes philosophers, people with religious backgrounds, AI engineers, politicians, and governmental institutes based on law regulations as a multi-profession group. Some interviewees think that the government is not able or the right candidate for developing such AI guidelines based on missing experts. Others say that it must start with a political action. As a requirement for the AI guidelines creators, it is highlighted that the actors will need to understand and consider actual problems and features of AI development in detail and not only generating abstract statements. Further input is that industry consortia or the AI system manufacturer should not take the lead as they might shape the AI guidelines in their own interest. One participant would like to follow public guidelines as a blueprint but wants to adapt it for internal company use. Private institutions or private organizations can develop AI guidelines in a faster way as they are already required. Another input from the interviewees was that from an economic perspective, the European Commission established industry association or the existing ethical review committee and that should take the lead. Another interviewee proposed an independent external third party like it is done for a financial audit or mandatory rules like within GDPR.

6.4.5   Advantages of AI System Certification

Interview participants can imagine that their company might certify their AI product or service as it can be helpful to commercialize their AI product or service. This requires a common standard that is also easy to implement. Implementation drivers can be requests from clients or customers, awareness of an additional value, marketing value, if it is mandatory forced, as an internal review tool, and a trusted and well-known AI certification label. Interviewees see the ethical AI certification coming anyway and describe the automotive industry as a leader. AI certifications and audits are more realistic to happen with the usage of augmented reality glasses. A mentioned advantage is that companies can react easier to a negative issue based on the finalized certification process, as it is seen as a preventive action. Further, it offers transparency to others and should include the certification of the data sets and the process itself as well.

6.4.6   Doubts About AI System Certification

Several interview participants do not certify their AI system based on the reason that it is not visible enough for the broader society or due to expected high certification costs. It can be considered if it would have a positive effect. Customers never asked for any ethical AI certification depending on either they do not know about it, or they do not care about it. Another concern about certification is that it can be difficult to measure as their companies' AI product or service is only a part of the whole final product.

6.5   Social Responsibility

The main category "Ethical and Social Aspects for Societies" and its subcategories "Ethical Aspects of AI Systems," "Environmental Impact," "Social Impact," "Societal Impact," and "Legal System Adaption" are referring directly to the fifth sub-question (cf. Sect. 1.1) to answer the central research question. The findings of the interviewee's answers generate insights into whether the management considers social responsibility and societal impact for their AI product or service development. Therefore, the answers will generate insights into the existing awareness by the management.

6.5.1 Ethical Aspects of AI Systems

One interviewee highlighted that the goal is not to maximize happiness, it is about to reduce the pain for large parts of the society. Another aspect is that AI systems might be unsafe and can harm and discriminate people by biased data algorithms or wrongly implemented incentivization. Further, it is mentioned that AI systems are based on software and that nonethical written software code can influence the behavior and ethical AI guidelines and discussions should be broadened to a general ethical coding topic. If society is not aware of the functionality and possibilities of AI, humans can misunderstand the technology, and fear and doubts can increase.

On the other side, AI system developers need to generate transparency in their AI systems to help others to understand the behavior. Another input is that from the interviewee's perspective, it is too early to discuss, but users should not be traceable, and the AI systems should refer to numbers regarding data privacy. As positive aspects, the usage of AI algorithms to identify biased data or predict use cases are mentioned. In medical context, a wrong decision or interpretation can lead to self-fulfilling prophecy and further diseases. The importance of data protection and the uncertainty about nontransparent processes where AI systems are making decisions for humans are highlighted. This includes transparent and explainable methods of how and from where the data sets have been collected.

6.5.2 Environmental Impact

Many interviewees' companies are not taking any environmental action, but one is asking partners for it. As an argument, it was mentioned that the company size is still too small for it, or it would be different if a huge number of servers combined with high financial cost would be used. An input is that it can be difficult to identify product, services, or third-party companies that are offering environmentally friendly or sustainable services. Projects about production process optimization often has as main or secondary goal to save energy based on cost optimization. As a positive impact, the change to buy green energy, reduce computational power, buy environmentally friendly products and services, or switch off energy consumption devices, if they are not needed, are mentioned. Participants think that it is still too early for a broad awareness about AI's environmental impact and companies are not taking this into consideration. It is mentioned that the management does not have a choice about switching to alternative cloud services as they are already in a strong relationship with the services from leading technology companies like Amazon. A positive input is to use crowdsourced pollution data, e.g., from cars to optimize and reduce traffic within cities.

6.5.3   Social Impact

One interviewee highlights the circumstance that people are using AI products and services in their daily life without being aware of it but are influenced by it. The participant describes that based on his/her work experience, management is afraid of job loss and to start gaining new knowledge by deeper research and they are not trying to change their company. AI systems can improve team efficiency but requires a broad technology understanding. Healthcare systems worldwide have difficulties serving the patients' demand based on limited manpower of professionals. AI systems can help humans in many countries get access to healthcare. In internal research projects, employees would not continue working on projects if these would generate job losses for others.

One company's goal is to reduce simple work tasks and to shape more meaningful jobs for the future. Reducing jobs by AI systems has been discussed within the company, but this did not happen within the AI system development company, it was instead transferred to their customers, where some people also got promoted. Management needs to decide carefully not to lose customer trust by doing unethical actions to fulfil cost savings, growth, and process optimization. An interviewee thinks that these decisions are not up to one company but are standardized within the industry and companies are part of it. Companies are not considering job losses based on AI system implementations, the focus is instead on doing the job tasks. This is seen as a competitive element, and it is not a social problem. The installation of robots can be done to assist humans as the robots are not able to replace people. Further, new jobs like maintaining and programming the robots are created. The interviewee is not sure if the consideration of job loss is a task of the management and the development of AI systems.

6.5.4   Societal Impact

Based on geographic areas and income, AI systems will have a strong impact, especially on low-paid workers and jobs. The interviewee thinks that companies listed on the stock exchange are focusing less on general societal improvements and (data) privacy is differently handled in Germany, the USA, and China and this influences the speed of AI system development. The interviewee's company in the medical business is evaluating the consequences of AI systems for the overall healthcare system. The banking sector is aiming for job reduction by AI systems and industry domains are focusing on optimization of production processes. The retail market is competing with big leaders like Amazon and AI systems can help them to stable or gain market share.

Companies are investing in AI systems to achieve cost savings, growth, and process optimization. AI robots can enable senior citizens with access to digital content. An interviewee said that the societal impact consequences of their AI system are not significant as all the user-specific data is not recorded by their system. Important factors are development speed, transparency of the decision-making process, and trustful AI algorithms. People underestimate the impact of ethics in technology and are mainly focused on the outcome and results. AI systems can fight against fake news and misinformation, and it requires awareness about abusing AI technologies for example to manipulate elections, affect regulations in a negative way, or within employee recruitment process.

6.5.5   Legal System Adaption

Many interviewees do not imagine the government as a leading law or guideline creator due to the pace of AI technology development versus law creation and approval. One proposal is to handle all AI issues in a newly created department. The legislation should act on a very general level to define high-level principles that describe what is allowed and what is prohibited without the need to constantly adapt the laws. The legal system should enforce that the AI systems are always explainable. Under the legislation, public institutions like the United Nations, the Global Court of Justice or work councils constituted of scientific, AI and legal experts can be installed. The focus on this level can be on analyzing the changes, identifying the key trends, and evaluating guidelines with a proactive agile approach.

Data is the main input for AI systems and therefore it needs to be ensured that quality outcomes require quality input data and people's trust in AI systems. Overall, it is welcomed to have a legal regulation for AI systems as what is existing for human actions, considering that people can write nonethical software codes. Therefore, the legal ethical framework should not only belong to AI, but it must also cover general software development principles. This will affect the job roles, requirements, and understandings of the legal and law sector, where more software and AI knowledge will be required. Further aspect is that future law regulation will also deal with historic data and crimes that happened years in the past as well as with predictions to estimate the future impact.

6.6   Responsibility of AI Systems

The main category "Ethical and Social Aspects of AI Systems" and its subcategories "Transparency," "Robustness," "Bias," "Privacy and Security," and "Accountability" are referring directly to the fifth sub-question (cf. Sect. 1.1) to answer the central research question. The findings of the interviewee's answers generate insights into whether management considers social aspects of their AI product or service. Therefore, the answers will generate insights into the existing awareness by the management.

6.6.1 Transparency

German companies want to understand how the AI system decides, but often the decision-making process is acting like a black box and others need to trust the outcome. Interviewees highlighted transparency as an important topic for decision-making and meaningful interaction, but there are following concerns about being transparent: others need to understand the AI technology to interpret the behavior correctly, too much information can confuse or lead to wrong self-interpretations, disclosure of core algorithms and intellectual property or commercial and competitive issues. To generate transparency is difficult, especially when it comes to deep neural networks, where it is not clear how easy it is to achieve transparency at all. Transparency in decision-making processes can be achieved by having a human in the loop to make decision together.

An interviewee's company is using explainable AI and another one has an internal group of employees who work on integrating privacy and explainable AI. Other customers have never asked for transparency and the AI system developer would not like to share deeper insights into their used technologies and algorithms. If nontransparency exists, it is difficult to understand if the outcome is, for example, biased. One interviewee's company developed a transparent data marketplace from the beginning to collect data and to build algorithms in a transparent and ethical way. The marketplace is auditable by any third party and will offer all their algorithms as open source in the future. To make their whole process transparent without disclosing their core intellectual property, they are using blockchain technology and encrypted training of algorithms to gain people's trust in their algorithms.

6.6.2 Robustness

Cultural differences like a stronger focus on robustness for German companies compared to Chinese ones are mentioned by an interviewee. Robustness is seen to prevent harming humans by keeping a human in the loop. Robustness is seen on the whole product or system level and is not dedicated to the AI components as it is difficult to isolate AI modules. Often industry security standards are implemented to harden the system with different security measurements and penetration tests. As a pre-requirement, it is mentioned that the model has been trained with the correct data sets and that it is also robust against false or wrong input data and configurations. Regarding the input data, it is unclear whether the AI system will use trustful and unbiased data or not. One mentioned approach to increase robustness is to use encryption for data and block-chain technology to make it accountable as well.

6.6.3   Bias

Technologies like facial recognition, where a high amount of bias is happening right now, should not be deployed in public applications. To reduce bias in AI systems it is proposed to increase employee diversity, verify if the model has been trained with unbiased and quality data, use AI algorithms to identify bias, implement collaboration and data sharing or assure transparency of AI algorithms. Biased AI applications can discriminate and harm people. Therefore, one of the interviewed companies has always had humans involved to increase the robustness against bias. They are aware that also humans can have a bias, but they rate this as a low negative impact factor for their medical knowledge database produced by humans. In the industry sector, biased sensor data can lead to, intended or not, more maintenance cycles. Interviewees are aware that their used data sets are or might be biased.

There were different opinions among the interviewees when it comes to bias if nonhuman produced data like machine data can have a bias, consumer data cannot be biased as it is the data of the customer, the understanding that every system has a bias or that data needs a bias that it is useful. The data bias requirement was mentioned with the example that an experienced lawyer will decide differently compared to a law study graduate. The interviewee calls this impact on the decision bias, and it is generally known as expertise. One interviewee prefers to use the term fairness instead of bias. In job application processes, biased data and algorithms can lead to unfair decisions based on gender or age. Another aspect is if the AI system will take similar or identical decisions as a human with the same input data. The Transfer Learning technology generated bias and was not diverse enough for the AI system development of an interviewee and depending on the context and usage, AI technology can generate different results.

6.6.4   Privacy and Security

There is a different international understanding of how to handle privacy and security within countries. A common future way can be to use Federated Learning technology, where private data stays on the device. Therefore, Europe can be a leader by pushing this development. Another approach is to separate personal from measurement data or to install the position of a dedicated data protection officer to protect and establish a good company reputation. One mentioned aspect is the ownership of the used data. This needs to be considered, but it is not always clear who really owns the data.

Other interviewees say that they do not have a privacy issue as they do not use personal data or use anonymized and pseudonymized data. Some mentioned to follow the GPPR rules as a guideline, restricting the AI systems from accessing sensitive data, data encryption, protecting the data by not exposing it to the team or application, or avoiding tracking on the personal user level. In that way, it is not analyzed what a single user is doing, instead they gain knowledge about what a population of users is doing. An AI algorithm understands input data sets as characters but does not understand that this is the personal data of humans. Overall, privacy needs to be secured so that people can trust AI systems.

6.6.5   Accountability

In the USA, many company board members or leaders are the ones who developed the software and have therefore strong technical knowledge, in comparison to other countries like Germany, where one or none of the top management members has data engineering as a background. Regarding AI accountability, companies do not have a clear strategy about how to proceed if a mistake will happen. During the interviews, different approaches were mentioned: the software developer should be accountable if they used the wrong data, the data science team, the overall AI developing company or the quality assurance department within the AI developing company.

Companies would differentiate by which component of the product or service would be affected and think that it is traceable to analyze the cause of a mistake. Prevention methods are testing and verifying the results with customers, develop a system for identifying failures and investigations, to reconsider how the user can override the AI system or the intended usage, and to integrate technologies like blockchain to make the whole process accountable.

6.7   Management Reflection

The findings of the interviewee's answers regarding this main category generate insights into an interesting aspect or something important for themselves and for their AI company. Discussing and exchanging ideas in workshops and interacting with others facilitate the achievement of impact that is created with new technologies and topics. The interviewees' company has strong principles but does not have a formal guideline or manifesto about AI principles. The interview created an awareness of the importance to have an AI guideline and as a result, the company is now planning to create one. Another important aspect was to speak about auditing and certifying an AI system regarding ethical principles. This was not considered earlier. Ethics of AI has many different aspects, it is not assigned to one discipline and requires multidisciplinary stakeholders. The discussion about the term's moral, ethics, and artificial intelligence. The new perspective on social impact and accountability, to reflect the own and other company's positions regarding AI ethics and to speak about regulations, which are not there by today, are all important topics and are required for the future.

6.8   Findings and Recommendations

The findings of the empirical research have been summarized in categories and the results will be discussed in this chapter. Referring to the central research question and its sub-questions, the discussion will be about the management considerations of ethical and moral aspects for their AI development. Further, the usage of ethical AI guidelines and their pros and cons will be answered. In the end, the motivation of the managers and their awareness about their influence on social impact with their AI product or service will be discussed. The empirical research generated more findings and new aspects, but in this discussion, the focus will be to answer the defined central research questions and will discuss AI from a social and nontechnical point of view.

The first sub-question of the central research topic is answering the management's understanding of their used definitions and terms. Based on the outcome of the literature study, there are many existing different definitions, and this question covers the perspective of the management. A common definition and understanding will make it easier to generate ethical AI guidelines and certifications on national and global levels. Moral was defined in different ways but represented a similar meaning compared to literature. Sometimes it was difficult for the management to find an exact distinction between moral and ethics. The term AI was often described by an example of the own AI product or service, but overall, like different literature definitions. Future AI ethics will increase complexity and will require perspectives of different professions. This may lead to a new diverse AI ethics discipline that will influence different existing professions like for example philosophy, psychology, law, software, and data engineering.

The second sub-question of the central research topic is about the prioritization of moral and ethics from the management perspective. The aim is to understand if managers are setting a focus on AI ethics and to understand their motivation behind AI development. Firstly, almost 80% of the interviewed managers rank the prioritization of AI ethics and moral as high. Secondly, the answers from the open questions identified an overlap of AI ethics, data ethics, and ethics as answers started by AI ethics and moved to ethical principles and company values. There are approaches to take the internal company values and ethical guidelines as a template to design ethics for their AI product or service. Often, ethical company principles are existing in an implicit way, and they are not written in a document. This does not guarantee that every employee knows them or even understands them in a similar way, especially if the company has multicultural employees. Another aspect is that companies can develop AI components that do not harm humans and their privacy, from the interviewed managers' perspective, but the final AI product or service, done by another company, may do. In the interviews, managers answered that in some cases they are not aware of their customers' final AI system, where their AI component is used.

Should ethical AI be considered at the main AI development company or on the supplier level as well? Besides this, AI systems can require further ethical principles that are not used within company values. Several managers mentioned that their younger employees would not work on unethical projects and that this is an important factor to keep and recruit employees. Therefore, the AI ethics topic is not only relevant to the product or service, but it also represents the company and its brand as well. When it comes to the motivation behind the foundation of an AI company, most of the interviewed managers mentioned customer, product (data), or curiosity as a driving force. Either the goal was to solve a customer problem and to increase customer experience or to predict future state by data analytics. Both cases, customer and product, are business oriented and it is not clear how strong AI ethics is represented by the final AI product or service.

The third sub-question of the central research topic is answering if there is an existing diversity in the AI development within the companies. Are technical IT people or other professions leading the AI development process? The interview results define most internal AI development stakeholders with a technical background in computer science, data science, software engineering, mathematics, statistics, and physics. The strong technical focus is not generating a diversity of different professions and current job offerings are mainly focusing on employees with technical AI skills. Diversity is one factor to reduce bias. Further involved professions are company managers, project managers, legal advisers, business developers, sales, and marketing managers and from the external position the customers.

Employees with a psychology background that does not necessarily require a university degree, are working in the field of user experience, in human interaction research projects or marketing. This research did not further analyze, if those stakeholders will have a direct influence on the ethical AI product or service development. It was mentioned that the external stakeholder customer considers AI ethics less than the AI system development company. This shows that the ethical responsibility lies with the AI system manufacturer side and therefore, the management of the AI company needs to understand and implement ethics for their AI systems. It has earlier been mentioned that the interviewed companies lack people with social studies background and the reason behind this has therefore been analyzed:

- The company is too small: It is though unclear how the size of a company is defined. Is it a question of headcount, revenue, or amount of sold units? When is a company too small or big enough to take social responsibility? Is it a prioritization between money and social interest?
- Philosophy does not match the current AI development: Does this mean that philosophical aspects never change or that AI systems are shaping the future of philosophy? Does one need to fit exactly into the other shape or are there flexibilities to combine and develop AI philosophy?
- It is not important enough for the business: That would mean that other tasks have a higher priority than AI ethics. It is unclear what kind of tasks are having a higher priority. Could it be that these tasks are financially related? Is it about for example generating more money or offering more jobs? Business is shaping the economy and influences social wealth, status, and societies. How are these topics prioritized by the management?
- It is not considered by the management: Are they not aware about it or does it have a low priority for them? If ethics is not considered for the development of AI systems, will future AI systems be ethical? Or will A(G)I systems develop their own ethical guidelines based on random data?

The fourth sub-question of the central research topic is answering if the management already knows and uses AI guidelines and principles within their company. Besides this, the managers were asked about their opinions about advantages and disadvantages of ethical AI guidelines, who should take the creator lead and their perspective about an ethical AI certification.

Some managers mentioned that they have a company value document that gives a guideline to employees and management. Those guidelines are having a universal character but are not explicitly referring to AI ethics. It is about a common frameset of company values that can be used as a basis for ethical AI guidelines. Most companies do not have any explicit ethical AI guidelines. The reason for that is based on arguments like that the AI product or service is not violating human rights, harming people or the management does not see any need and rated the topic as less important. What is still unclear is the ethical usage of the overall AI system by the customer.

Will ethical aspects change during future product developments and market adaptions? Will the company create an ethical AI guideline later, or is it then too late? The company's argument also demonstrates that actions would be taken or considered only if the AI system might harm in any way. There is no consideration to take any action as a prevention or a way to change the future development direction in a positive way, if the AI system might harm in any way. It is not seen as a prevention or a way to change the future development direction in a positive way. During the interviews, some managers realized the benefits and the need to create an internal ethical AI guideline for the company, as it is easy to implement into the existing business and the employees will also accept and live the principles.

Another ongoing discussion is about the core principles of ethical AI guidelines. Therefore, one approach during the empirical research was to verify the results of Rothenberger et al. (2019) from the management perspective. The task for the managers was to bring six guideline proposals in order depending on their importance to them. "Responsibility" was ranked as the most important and "Protection of Data Privacy" as the second highest in both independent research studies. As the top two ranks are similar, those two principles should be the focus points in the development of general ethical AI guidelines. Most of the managers described the creators of ethical AI guidelines as representatives of the society from every country and nation with a good understanding and interest in AI technology.

This includes a diverse multi-profession group of philosophers, people with religious backgrounds, AI engineers, politicians, and governmental institutions. The start of the AI guideline development should be based on law regulations but should also be flexible to update the guidelines regularly. Many managers do not see the government in the lead based on missing experts and AI knowledge, but some say that the start of developing ethical AI guidelines can be triggered by a political action. Some also see the European Commission, industry association, or the existing ethical review committee in the lead. As there is already work ongoing, for example, by the European Commission, which most of the managers were not aware of, it is unclear whether the communication of those institutions does not reach the companies or if the managers set their focus on completely different directions.

If all stakeholders (whole society) will not come together and align about one AI guideline, we will end up with several independently created ethical AI guidelines by different actors without any implementation. Another management input was the consideration to use a public guideline as a blueprint for customization within the own company. An implementation can be ensured by rulesets like financial audits or GDPR implementations. Besides the amount and definition of the ethical AI principles, the willingness for a (global) implementation will play an important success factor.

Therefore, an ethical AI certification system offers transparency to others and the managers see the certification as a preventive action that can also be used as an internal review tool. Besides, it can help their AI product or service to commercialize, generates additional marketing value and trust. Some managers can imagine themselves certifying their AI product or service if there is a common standard and if it is easy to implement. Especially the last two points are of a high importance to the management. What was not part of the research is the definition of a common standard for management. Is this a national, European, or global standard? If it is only valid nationally, will there be any problem with international sales and distribution? Will international markets require different certifications?

Maybe it would be possible to adapt from existing certification processes like the American FDA or the European CE certification. Another open point for ethical AI certification is how to handle the AI system certification if the manufacturer is implementing several different AI components from third-party suppliers. Is it possible, from an ethical point of view, to certify all AI components beforehand or has this to be done by the seller? Management concern is about high certification costs and that the certification is not broadly known within societies. If customers would ask for it, most managers can consider an ethical AI certification. Based on the interview results, the manager's customers never asked for any ethical AI certification or explanation and therefore the companies are not certifying their AI system. Nevertheless, the management sees the AI certification coming, and this will be most likely be driven by the automotive industry leaders.

The fifth sub-question of the central research topic is answering what kind of social responsibility the management has, regarding their AI product or service. Therefore, the first part includes answers about ethical aspects of AI systems, environmental, social, and societal impacts, ending with answers about legal system adaptions. A manager describes an ethical aspect of AI systems by focusing on the goal not to maximize human happiness, instead to focus on reducing the pain for large parts of society. As an example, investment in AI medical or agriculture applications instead of AI weapon systems can be mentioned. Why is humanity not taking care of the poorer ones instead of increasing its own profit?

Would it be more difficult for nations to live together if they all would have a similar wealth level? AI technology can have the power to make an immense impact and change the future world order. As a concern, it is mentioned that AI systems can be unsafe and harm or discriminate people based on biased data algorithms. But why might this be the case? Are those AI systems not developed by AI companies and their management decisions? Research shows that there are concerns about the mentioned arguments and despite that almost all the interviewed companies are still not using any ethical AI guidelines for their own products or services. It is mentioned that nonethical written software code can influence the AI system behavior negatively but who writes the software codes?

Software systems can only be as ethical as their developers decide to and A(G)I systems might learn on top of it. It might also not be clear on what kind of understanding AGI systems will develop their ethical guidelines and moral decision making. Will an AGI decide to take care of the poorest humans, or will it support the strongest nation to get settled on other planets? Therefore, an early ethical AI standard will generate a long-term impact on humans, and it might be a general topic for developing software systems. Another aspect is that humans need to understand AI technology to lose their fear or to change the picture of a Terminator robot. Therefore, people from all age groups will need to learn, interact with AI, and create their own opinions to shape a responsible future. AI developing companies, also if they do not have direct end-consumer contact, are playing an important role in the whole ecosystem.

Many companies of the interviewed management do not take environmental actions for their AI development based on arguments that the company is still too small, and it would make a difference if the company would operate a high number of servers to reduce costs. Therefore, the decision-making argument to switch to alternative environmentally friendly products and services is based on costs. However, if a company would now decides to change, how would it be possible to verify an environmentally friendly status of an operator? For the management, it is difficult to identify these companies, as many are using environmentally friendly marketing slogans. Another argument was the already existing dependency on the main supplier, for example, for cloud services, where it is difficult and expensive to change the company. For the management, it is still too early for a broad awareness about the environmental impact of AI systems, and they believe that companies do not care about it. If that is the case, could a government regulation help to take care of the planet? Independently of the company size, everyone contributes to the planet's future.

A frequent discussion topic regarding social impact of AI is the way how humans will handle job loss and job changes in the future. Overall, this topic is not new and started already before automation of production processes during industrialization. However, the factor speed is different this time. Changes are happening faster, and humans will need to develop even more flexibility in the future. Management is aware of possible job losses, but on the other hand, there are also concerns about learning new technologies and thereafter to adapt the companies. Management is focusing on cost savings, growth, process optimization, and sees the social impact of AI systems as a competitive element and not as a social problem. Arguments like the generation of new jobs and human-assistive products were mentioned. Are those new jobs only for new generations or will it be possible to migrate older people or other professions as well? Will an unemployed truck driver or a bank employee find a future purpose? The management is not certain if the responsibility of possible future job loss is a task for the company management. Another approach can be that the government decides and limits the research and development of future (AI) technologies to special industry sectors like, for example, healthcare.

Focusing on the societal impact of AI systems, the managers point out how different countries like, for example, Germany, the USA, and China handle data privacy and data sets as input data for AI algorithms. This generates use cases that are not allowed or limited in other countries and leads to fast-developing AI nations in less restricted countries. In contrast, a high data privacy regulation can protect human privacy and diversity. Overall, AI systems will influence global society in many sectors like healthcare, retail market, elderly care, education system, and jobs. Some interviewed managers described their AI product or service as generating less impact on societies. Therefore, AI systems should be categorized into high- and low influencers based on several criteria.

Many interviewed managers cannot imagine the government as a leader in AI law and guidelines, as they think that the pace of AI technology development is faster than their decision-making process. Managers consider the government as a high-level institution to create a general framework that will be further developed and updated by a group of public institutions, work councils, AI, and legal experts as representatives of the society with diverse professions. It is proposed to include data and software developing ethics as a part of AI ethics. The quality level of input data is influencing the behavior of AI systems. Management is not always able to verify the input data sets and therefore trust the data without any quality and bias verification. The generated user and machine data from the past years, might be used in current and future AI systems. Therefore, the law and legal sector will need to adapt historic data analytics as well for predictions of the future.

Managers define transparency of AI systems as an important topic preventing ending up with a black box that nobody can understand without any possibility to interpret its results or behavior. On the other hand, managers have concerns about being transparent when it comes to their own AI product or service. Due to competitive-related issues, companies want to protect their intellectual property and are not willing to be completely open. Some managers mentioned that their customers also have not asked for transparency details, but others say that especially German companies, above all, want to know how the AI system decides. This may show a paradoxical situation that companies do not want to share but require transparency at the same time. Does this need a governmental regulation, or will companies handle this by themselves? One interviewed company found a way to combine other technologies like blockchain encrypted training of algorithms to create a completely transparent process. That could probably be an approach for other use cases as well.

Robustness is defined differently within cultures and countries, but overall, it is seen as a prevention of harming humans. This should not only focus on the AI component and should instead include the whole product or service. Companies are increasing robustness of their AI system by implementing industry security standards, doing penetration tests, and analyzing data sets. As AI systems might control important systems and infrastructure facilities like power plants, the awareness about putting effort on increasing product or service robustness is existing.

One approach to reduce bias in AI systems is to generate employee diversity by profession, age, gender, cultural background, etc. and to use quality input data for AI algorithm training. Since humans might also have a bias, it is difficult to generate completely non-biased AI systems. Interviewed managers are aware that their used data sets are or might be biased. If this happens already in an early stage and complexity is increasing, how will the AI system look like in the future? Will it identify bias by itself, or might some bias be a kind of new normal? The future discussion would also require a classification of the bias types as not every bias is necessarily negative. Depending on the situation and circumstances a bias can be seen as positive by humans like, for example, a belief to trust most other people, but will an AI system understand the context for making the decision?

Another mentioned aspect is to use the term fairness instead of bias. However, it is important to keep in mind that fairness is not a uniquely defined term based on culture, socialization, and experience the term fairness or fair stands for completely different understandings and actions and should therefore not be used within this setting. Current AI systems are good at finding patterns, so it would be a positive use case to train and install AI algorithms to identify biased data.

Privacy and (data) security is handled differently within countries and Europe might have a leading position regarding data protection. The interviewed managers from European countries do have a strong awareness about data privacy based on different regulations like the GDPR. An aspect to highlight is that an AI algorithm will understand input data sets as characters, but it does not understand if this is a personal data from humans or not. This needs to be defined beforehand to gain people's trust in AI systems. Management thinks that in big European companies the C-level board needs more people with a strong technical background in computer science and data engineering to break into a digital (AI) future. Regarding accountability, the management can consider different approaches, which are coming from software development processes, to how to handle AI system failures. It is not clear if the accountability of AI systems will need additional procedures since this will involve more stakeholders in the future.

The first research part of this book shows how complex and still partly unanswered the topic about ethics of AI from a management perspective is. Besides the technological complexity, it also affects other disciplines, professions, and nations that need to cooperate, shape, and implement global frameworks and standards. Further digital technologies and trends, partly driven by AI, will generate an immense impact on consumer behavior, economy, societies, and other sectors. Industry sectors, for example, the retail market will need to positively transform itself into a successful combination of online and offline services. Standard food deliveries can be automated and can save time for consumers, but the in-shop experience, e.g., fresh meat or vegetables can generate new business value via facial recognition, AI-driven customer profiles of habits, connected devices like shopping baskets, fridges, or robotic assistants. Is this the future that makes everything better for all humans?

AI will change the education system in the future. Instead of teaching a full classroom, digital AI assistants can analyze each pupil's knowledge level and can generate personally adapted courses for each level. Wrong usage of digital surveillance tools might lead to generations of humans without their own opinions and creativity. Therefore, ethical lessons (independent from religion) should be included in an early stage of school to open possibilities for discussion, diversity, and personal development.

Growing unemployment rates and heavy job losses are frequently discussed topics. Will be a universal basic income (UBI) the solution? Experiences from the past (invention of steam machine and electricity) show that humanity and economy have evolved and new ages were born. Will this happen in a positive way with AI again? Will humans consider opportunities for the immense increasing development speed of new technologies and job deskilling this time? Will the world be full of software engineers and data scientists in the future? What about the others? Digitalization is a powerful transformation for global players and monopolists to expand their market power even more. Will there be opportunities for small businesses in the future or will the price and the delivery speed (of the monopolists) be the only valid success factors? What kind of moral and ethical understanding will future humanity have?

This research aim was to analyze the management perspective and awareness about ethics in AI. The management perspective has been chosen as the company management drives the development and implementation of AI products and services. The interviews generate new perspectives and information for both sides and show that there are still many undefined and nonregulated issues. This multidisciplinary topic has immense potential to change the future of humanity and therefore all people should be involved in the development and usage of AI.

Reference

Rothenberger, L., Fabian, B., & Arunov, E. (2019). Relevance of ethical guide-lines for artificial intelligence – A survey and evaluation. In *Proceedings of the 27th European Conference on Information Systems (ECIS)*, Stockholm & Uppsala, June 8–14, 2019. ISBN: 978-1-7336325-0-8 Research-in-Progress Papers. https://aisel.aisnet.org/ecis2019_rip/26

7

TAII Framework

The evolution of AI systems can be seen from two different perspectives: as a negative strategy trying to prevent disasters and keep the AI system fulfilling its originally defined purpose or as a positive strategy that enriches the AI system's benefits to humanity (Boddington, 2021). The creation and usage of data for developing AI systems in an ethical manner evolve the need for data regulations in a connected digital world. The implementation needs to be done carefully around tension between technology innovation and protection of privacy (Wachter, 2019). AI systems should not undermine the aims of the European General Data Protection Regulation (GDPR) (European Commission, 2021a). Instead, the GDPR should be seen as an enabler for implementing trustworthy AI systems. To govern AI systems can be a challenging task in the field of tension between different internal and external stakeholders, competition, and markets. Regulations should protect the fundamental rights of humans, but they can also generate tension regarding international competition and innovation. Therefore, a successful set of AI guidelines requires an equal balance between technology and innovation, politics and state, economy and market, and humans and society with their environment (World Economic Forum, 2019).

© The Author(s), under exclusive license to Springer Nature Switzerland AG 2023
J. Baker-Brunnbauer, *Trustworthy Artificial Intelligence Implementation*, Business Guides on the Go, https://doi.org/10.1007/978-3-031-18275-4_7

The development of Trustworthy Artificial Intelligence (TAI) systems creates the need for practical tools and guidelines to initiate the implementation of AI ethics within companies for their products and services. This chapter introduces the Trustworthy Artificial Intelligence Implementation (TAII) Framework to tackle this need. As such, this research aims to fill a literature gap for management guidance to initiate trustworthy AI implementation while analyzing ethical inconsistencies and dependencies for the planned AI system. Whereas other research on trustworthy AI (see Sect. 7.1) has primarily focused on the definition and implementation of ethical principles, the TAII Framework* is comparatively unique given that it considers the holistic perspective of developing and implementing trustworthy AI systems within organizations. Instead of starting directly with the implementation of ethical principles for the development of AI systems, the TAII Framework offers management guidance to initiate trustworthy AI implementation by starting with the analysis of ethical inconsistencies and dependencies for their planned AI system. The TAII Framework provides guidance for the involved stakeholders and considers these dependencies: corporate values, business models, and common good. Section 7.1 describes the background of the work, Sect. 7.2 explains the trustworthy AI approach of the European Commission, Sect. 7.3 presents the TAII Framework, Sect. 7.5 describe transfer challenges, and finally Sect. 7.6 offers using the TAII Framework Canvas a productive way how to start interactively implementing TAI. Trustworthy AI can generate major improvements in the areas of humans and society, private and public sectors, research and academia, availability of data and infrastructure, skills and education, governance and regulation, and funding and investment (European Commission AI HLEG, 2019a).

7.1   Introduction

An AI system should be seen as a socio-technical system whose impact is not only based on its design. Instead, the system should consider its broader environment, including purpose, training data, functionality and accuracy, scale of deployment, and the broader organizational, societal, and legal context (Council of Europe, 2020a). Data science opens new ethical challenges in different research areas: the ethics of data, the ethics of algorithms, and the ethics of practices (Floridi & Taddeo, 2016). To prevent ethical inconsistencies during and after the development of AI systems, the implementation of AI ethics should be accomplished with the support of a multidisciplinary meta-perspective by the key stakeholders. Starting with the translation of existing ethical principles does not lead to a common good solution for all as it focuses on already designated areas. For example, the use of an AI system to optimize animal farming may save costs and increase output but should also question systemic relationships to animal rights, diseases, environmental aspects, and dignity. AI ethics is a part of the ethical operation of a company. The existence of ethical guidelines is not a guarantee of utilization and strongly grounded principles require legal mechanisms for implementation (Hagendorff, 2020). AI regulation approaches that take the social contract into account are ranked among the most open ones to interact with society in coproduction with the government (Delipetrev et al., 2020). Research to address the multidisciplinary topic of AI ethics to embed political and societal contexts (Delipetrev et al., 2020) confirms that the answer to ethics implications of AI technologies requires a mix of law, design, and education (Calo, 2011). Besides making laws, the importance of discussions with society and academia to gain constant feedback is highlighted (Delipetrev et al., 2020).

More than 80 AI ethics initiatives published ethical principles and guidelines for AI system development and deployment (Hagendorff, 2020; Mittelstadt, 2019; Floridi et al., 2018; Twomey & Martin, 2020; KI Strategie Deutschland, 2020; European Union Agency for Fundamental Rights, 2020; Hickok, 2021; Cihon et al., 2020; Ryan & Stahl, 2021; Thiebes et al., 2021). Many initiatives envision to translate ethical high-level principles and abstract requirements, for example, fairness, transparency, or accountability, into mid- or low-level design requirements (Mittelstadt, 2019). The development of AI systems does not have empirically proven methods for translating principles into practical implementation (Mittelstadt, 2019). Needs and norms cannot be derived directly from mid-level design requirements without accounting for technology, context, application, and local norm elements. This requires normative decisions and the identification of coherences between principles, norms, and facts at each stage of the translation (Mittelstadt, 2019). Therefore, AI ethics implementation has some challenges ahead, and a common alignment of some high-level principles is only a first small step as shared principles are no guarantee for a trustworthy AI system implementation. Most of the tools and methods for implementing ethical principles lack usability and do not provide enough practical support (Morley et al., 2019; Vakkuri et al., 2019). To implement AI ethics with a top-down approach (from general legal regulations to AI system developers) is more difficult than a bottom-up approach that starts with the requirements and settings within specific use cases and applications (Mittelstadt, 2019). Empirical multidisciplinary bottom-up research might increase the speed of AI ethics implementation as it focuses directly on the needs and challenges of AI system developers. Besides giving attention to AI ethics on the development and deployment level, companies need to broaden their focus to the organizational ethics perspective. As AI engineers and developers will be constrained by their employers, AI ethics need to be aligned at the top levels of organization. Research shows a big translation, implementation, and accountability gap between practical transfer and guidelines of ethical principles for AI system developers (Shklovski et al., 2021; Baker-Brunnbauer, 2021a). This requires either additional skills for engineers or additional resources such as an "ethics-officer-in-charge." The TAII Framework supports the management of AI system developing companies to take actively the above-mentioned issues into account.

The implementation of AI ethics differs depending on the used technology, context, and risk level of AI systems. The European Commission proposed six requirements for high-risk AI systems: clear liability and safety rules, information on the nature and purpose of an AI system, robustness and accuracy of AI systems, human oversight, quality of training datasets, and the keeping of records and data (European Commission, 2020a). In 2021, the European Commission released the regulatory framework proposal on AI (European Commission, 2021b) that classifies AI applications into four risk levels: minimal, limited, high, and unacceptable risk. High-risk AI systems will need to fulfil more requirements than others by undergoing a conformity assessment to reach registration in the European Union database and to achieve the conformity declaration and CE marking (European Commission, 2021c). The TAII Framework supports the management of companies to develop trustworthy AI systems within each risk level.

7.2   Trustworthy AI

The European approach to trustworthy AI covers an ecosystem of excellence along the value chain from innovation and research to creating acceleration funding. Depending on the risk classification, AI systems that are developed or deployed within the European Union will need to fulfil the upcoming regulations (European Commission, 2021b). Therefore, the TAII Framework orientates on the TAI approach of the European Commission but can be adapted and used with different regulations and ethical principles. Furthermore, trustworthy AI will be shaped within an ecosystem of trust based on European fundamental rights and rules.

This will give users the trust and confidence to use AI systems (European Commission, 2020a). The implementation of trustworthy AI will gain public trust, clear responsibility, and enable "dual advantage" (Floridi et al., 2018). Acceptance by the public and adoption of AI systems will be successful if the usage of AI technologies is seen as a low risk and has meaningful areas of application (Floridi et al., 2018). The Sectoral Considerations on the Policy and Investment Recommendation for Trustworthy AI from the European Commission recommends a close collaboration between the industry and innovation ecosystems for research and transfer of trustworthy AI systems from ideation to rapid testing to deployment (European Commission AI HLEG, 2019b). Furthermore, it recommends proceeding within an open innovation culture within multidisciplinary research teams. The TAII Framework orientates on the European Commission's Ethics Guidelines for Trustworthy AI (European Commission AI HLEG, 2019c), but it is adaptable for others. The understanding of the European Commission is that trustworthy AI describes a human-centric and trustful development of AI systems to maximize the AI system benefits and minimize their risks (European Commission AI HLEG, 2019c).

To generate a common understanding of the main terms within the organization, the author recommends using a unique definition like the one from the AI High-Level Expert Group (European Commission AI HLEG, 2019d) which describes trustworthy AI with three components (lawful, ethical, and robust) that should be aligned through the whole AI system's life cycle (European Commission AI HLEG, 2019c). The implementation of trustworthy AI should be an agile and continuous cycle during the whole AI system's life cycle and covers technical (architecture for trustworthy AI, ethics, and rule of law by design, etc.) and non-technical (regulation, code of conduct, standardization, certification, education, and awareness to foster an ethical mindset, stakeholder participation and social dialogue, diversity, inclusive design teams, etc.) methods (European Commission AI HLEG, 2019c).

7.3   Overview of the TAII Framework

The development of trustworthy AI systems creates the need for practical tools and guidelines to initiate the implementation of AI ethics within companies for their products and services. Whereas other research on trustworthy AI (see Sect. 7.1) has primarily focused on the definition and implementation of ethical principles, the Trustworthy Artificial Intelligence Implementation (TAII) Framework considers the holistic perspective of developing and implementing trustworthy AI systems. The TAII Framework* has been developed by the author and offers a management guidance to initiate trustworthy AI implementation by starting with the analysis of ethical inconsistencies and dependencies for their (planned) AI system along the value chain. The TAII Framework contains 12 steps that will be continuously passed through during the whole AI system's life cycle.

The starting point for the implementation of trustworthy AI is the creation of an AI system brief overview (Fig. 7.1). This document describes the purpose, use case, and used input data of the AI system. The used and planned source data needs to be defined precisely as it is difficult

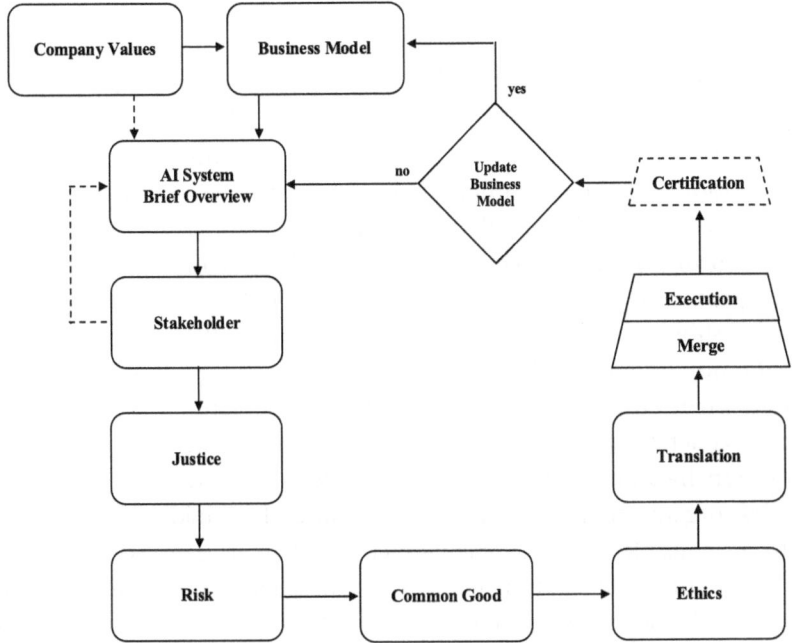

Fig. 7.1   Iteration of the TAII Framework

to implement AI ethics when the training data consists of biased data (Brandon, 2021). The TAII Framework is iterative and refers to the whole AI system's life cycle. This has the benefit that the first iteration does not require too many details to initiate the AI ethics implementation. Within the following iterations, more details need to be added to clarify possible misunderstandings and to sharpen the picture. The defined and aligned company values will be used for the development of the company's business model as well as to build a solid ground for the AI system brief overview.

Implementing AI ethics should correlate with a company culture that is based on ethics, morals, and values. To focus only on the implementation of AI ethics may cause contradictions with other company decisions. Therefore, it is recommended to see AI ethics as a part of the company values, morals, and adjacent ethics areas (Lauer, 2021). Otherwise, contradictions arise, such as implementing ethical principles for an AI system but buying raw material that was produced by disregarding fundamental human rights. Additionally, it is unclear how a company that acts unethically may (or based on regulations must) implement AI ethics without distorting reality. Company management should consent, document, communicate, and transfer the values to the company ecosystem (value proposition, stakeholder, supply chain, production, etc.) and establish an internal ethics board that will lead the TAII.

To ensure business success, explore the AI system's ecosystem, and strengthen management's commitment to the TAII, a visualized business model should exist from the beginning (Baker-Brunnbauer, 2019). Facilitation by independent externals supports making decisions in situations where ethical principles and values are conflicting with each other. The next action is to define and to document the stakeholders. This lists all internally and externally involved people, groups, departments, companies, organizations, institutions, etc. and each should name a specifically responsible persona. Stakeholders need to be categorized by role (contributor, responsible person, decision maker, accountable person, supplier, developer, compliance person, deployer, user, governor, auditor, etc.). Next, it is necessary to justify existing regulations and standards for the specific AI system within the area of application. If those exist, the legal requirements and limitations need to be considered and implemented. Different proposals exist to apply governance models for technological development and law formulation processes (Delipetrev et al., 2020; European Parliamentary Research Service, 2020; DKE German Commission for Electrical, Electronic & Information Technologies of DIN and VDE, 2020; Ben-Israel et al., 2020) as well as the Coordination Plan on Artificial Intelligence of the European Commission (2018) to safeguard a high standard of transparency, respect for democratic values, and legitimacy. It is the implementer's responsibility to be compliant with legal requirements and therefore this framework is not liable for the outcome.

The risk assessment of the ethical impact of the AI system is recommended and requires different measures depending on the risk level. Existing legal requirements for the application area and different assessment methods (Ben-Israel et al., 2020; Mantelero, 2018; Artificial Intelligence Ethics Impact Group AIEI Group, 2020; Krafft & Zweig,

2019) can be chosen to classify risk and social impact. The risk assessment should consider the AI system's potential harm and the affected human groups based on the unintended results of the AI system. Ethical risk should not be unobservable or unquantifiable (European Parliamentary Research Service, 2020). The rating is often defined between a low and a maximum impact level. AI systems with a high impact level are called high-risk applications in literature (European Commission, 2020a, 2021b; Council of Europe, 2020a, 2020b). Evaluating the risk in a two-dimensional matrix generates more clarity as the AI system may have different risk levels depending on the areas of application (Artificial Intelligence Ethics Impact Group AIEI Group, 2020). An industrial application may be classified as an ethical low-risk level, but the AI system may strongly affect parameters like energy consumption or job replacement.

The systemic assessment of the AI system in compliance with the common good via the Sustainable Development Goals (SDGs) (United Nations Sustainable Development Goals, n.d.) and the Universal Declaration of Human Rights (UDHR) (United Nations Universal Declaration of Human Rights, n.d.) will analyze dependencies and patterns within the organizational ecosystem (Systemic Society, n.d.). The AI system definition from the Council of Europe says the AI system should include its broader environment and its organizational, societal, and legal context offers a good basis to explore the AI system's broader impact on the SDGs and UDHR (Council of Europe, 2020a, 2020b). During the assessment of the SDGs, organizations will identify to which goals their AI system will contribute in a positive or negative way. Depending on the application or used technology, some goals are possibly not fitting for the AI system. Those can be skipped but the evaluation of the AI system's impact on the SDGs should be assessed critically. To improve transparency, the documentation of all answers and future iterations of the TAII Framework is recommended, which also generates new input. The reflection of the AI system with the 17 SGDs will broaden perspective, demonstrate the systemic dependencies of resources that should be used in a sustainable way, and question the connections for possible negative interference regarding the UDHR. AI ethics is only a part of how the company and its products and services will contribute to the environment and humanity.

The next step generates the list of ethical requirements and guidelines. Many AI ethics guidelines are already developed (Fjeld et al., 2020; Hagendorff, 2020; Jobin et al., 2019) and the AI system developer needs to align which ones fit best based on different factors such as (inter-) national regulations, legal requirements, field of application, and standardization. Within this research, the European Commission's approach to achieving trustworthy AI systems (European Commission, 2020a; European Commission AI HLEG, 2019c) has been chosen. This implies four principles and seven key requirements. The requirements are as follows: human agency and oversight; robustness and safety; privacy and data governance; transparency; diversity, non-discrimination, and fairness; societal and environmental well-being; and accountability (European Commission, 2020a; European Commission AI HLEG, 2019b, 2019c). The defined ethical principles and requirements need to be appropriately translated and transferred to the AI system ecosystem. This starts with the mapping of the previously defined principles with the application-related ecosystem. It is not effective to use a one-size-fits-all approach since, for example, the requirement "transparency" will have different aspects depending on the use-case, application, and context. Transparency can be understood in different ways, for example, to publish a short statement about the AI system's algorithm (communication), to make the whole source code of the AI system's algorithm public (decision-making process), or to achieve a certification that states the fulfilment of some requirements for the AI system's algorithm (comply standards). For the first iterations, it helps to follow checklist questions (European Commission AI HLEG, 2020) and to implement a specific mapping method in a more progressed iteration. This can be a method such as Values, Criteria, Indicators, Observables (VCIO) (Artificial Intelligence Ethics Impact Group AIEI Group, 2020) Value Sensitive Design (VSD) (Umbrello & van de Poel, 2021), Design for Values (Dignum, 2019), Value Sensitive Software Development (VSSD) (Aldewereld et al., 2015), data-driven research framework (DaRe4TAI) (Thiebes et al., 2021), applied ethical AI typology (Morley et al., 2019), and Artificial Intelligence Regulation (AIR) (Hagendorff, 2020). Companies will need to take both actions and self-responsibility to make the best-fitting decisions for the transfer of ethical principles.

After the mapping of the ethical principles, the merge of the previously assessed input factors (AI system brief overview, stakeholder, justice, risk, common good, ethics) starts. The goal of the merge is to define the current state, to visualize dependencies, and to plan the next tasks for continuous improvement. The execution verifies, tests, and implements the results. During the TAII all considerations and actions should be documented, and the responsibility of tasks must be clarified to achieve transparency. The execution of a mandatory (cf. AI high-risk applications (European Commission, 2021b)) or optional certification or safety assessment for the AI system will increase transparency and trust. Different institutions are working on certification standards (European Commission, 2020a; Council of Europe, 2020a, 2020b; Cremers et al., 2019; DKE German Commission for Electrical, Electronic & Information Technologies of DIN and VDE, 2020; Braband & Schäbe, 2020). As the AI system may change during every TAII iteration, the responsible stakeholder for the TAI implementation needs to reconsider if a new or recertification (assessment) is required or necessary.

After passing through the last step "certification" of the TAII Framework (Fig. 7.2), the next iteration starts with an update of the AI system brief

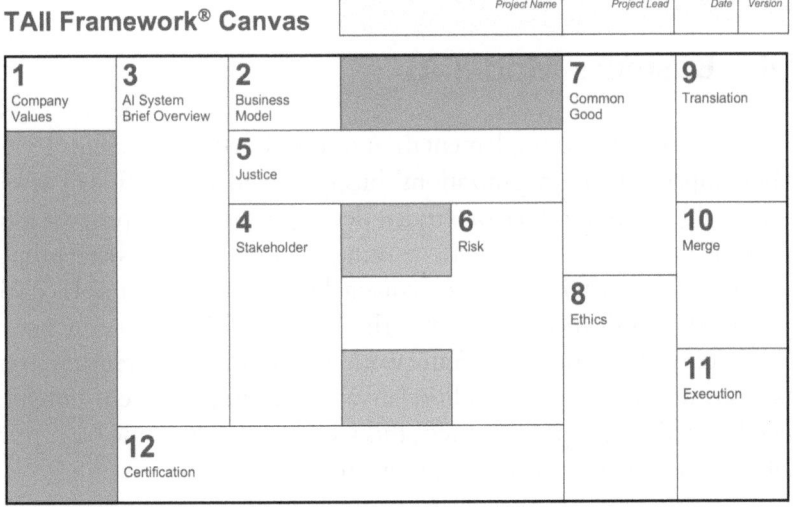

Fig. 7.2 TAII Framework Canvas

overview or if necessary, the business model. Some parameters will be changed or updated within future iterations as product, service, market, and society are evolving. The iterations are endless as long as the AI system's life cycle reaches its end. The area of application, risk assessment, maturity level of the AI system, and external factors (technology, market, society, etc.) will influence the speed and frequency of the iteration interval that needs to be defined by the AI system developer or deployer. The TAII Framework and its 12 steps can be used for existing and for future AI systems during the system's whole life cycle. Nevertheless, the success of the TAII is based on the defined priority given to AI ethics, planned and allocatable resources, and the commitment of all core stakeholders to implement and regularly pass through the TAII Framework. To improve transparency, a documentation of all answers and future iterations of the TAII Framework is recommended by the author. The author proposes to apply the TAII Framework to all risk levels of AI systems, including low-risk classification with less restrictive actions. Skipping TAII for low-risk AI systems should not be considered as it may oversee broader social impacts and dependencies. AI ethics is a part of the ethical behavior of the organization and influences its contribution to the environment and society.

7.4   Business Model AI

The development and implementation of trustworthy AI systems have a major impact on the organizations' business model on different levels: This might affect product or service development (value proposition), operation of the organization, existing and new processes, human resources and recruitment, decision-making processes, supply chain impacts, the profit mechanism, etc. The "Business Model" is only one of the 12 steps within the TAII Framework, but as it is important for organizations to reflect on the additional value and improvement potential outside the value proposition view, this section provides some basic principles of business models for management.

Business Model

A business model describes the picture of how an organization creates and captures value. A company's business model (BM) becomes specific by describing it in four dimensions (Gassmann et al., 2014):

- Customer (Who?)
- Value proposition (What?)
- Value chain (How?)
- Profit mechanism (Why?)

A BM must contain functional business logic and matching categories to generate success. It needs a holistic approach and a unique position and value proposition. A business model contains five components: Market position, product logic, value creation, marketing logic, and profit strategy (Matzler et al., 2016). Osterwalder and Pigneur (2010) write that a business model describes the rationale of how an organization creates, delivers, and captures value. A business model can best be described through nine basic building blocks that show the logic of how a company intends to make money. The nine blocks cover the four main areas of a business: customers, offer, infrastructure, and financial viability.

Clark et al. (2012) define a business model as the logic by which an enterprise sustains itself financially. It is the logic by which an enterprise earns its livelihood. Lewis refers to the term business model as a term of art and describes it with how it is planned to make money (Ovans, 2015). Peter Drucker describes business model as the assumptions about what a company gets paid for (Ovans, 2015). Others (Shafer et al., 2005) define a business model as a representation of a company's elemental core logic and strategic choices for creating and capturing value within a value network. This definition includes four key terms: strategic choices, creating value, capturing value, and the value network. Johnson et al. (2008) define a business model as a construct of four coherent elements (customer value proposition, profit formula, key resources, and key processes), which create and deliver value. These four elements are building the base for any business.

Business Model Innovation

A business model provides a holistic picture of how a company creates and captures value by defining the Who, the What, the How, and the Why of a business. A business model innovation (BMI) is the requirement to modify at least two of these four dimensions (Gassmann et al., 2014). The effort of solely innovating the value proposition would merely result in a product innovation. Today, it is best to assume that most business models will have a short lifetime, including the successful ones. As it is a substantial investment that a corporate make to create or innovate a business model, the enterprise should extend the lifetime through continuous management and evolution as long as possible (Osterwalder & Pigneur, 2010).

Lindgardt et al. (2009) describe BMI when two or more elements of a business model are reinvented to deliver value in a new way. It is important to distinguish BMI from product, service, or a technology innovation. Others write that any business model is essentially a set of key decisions that collectively determine how a business earns its revenue, incurs its costs, and manages its risks (Girotra & Netessine, 2014). Innovations are considered in the model as changes to those decisions: *what* your offerings will be, *when* decisions are made, *who* makes them, and *why*. Successful changes along these dimensions improve the company's combination of revenue, costs, and risks.

Business Model Generation

Previous chapter lists an extract overview about the different definitions of business model and business model innovation. As there are different definitions, there are different approaches to generate a business model. The design of a business model describes the covered dimensions of the business. Like the definitions of business model, there is no clarity or common agreement about which dimensions a business model should contain (Frankenberger et al., 2013). Therefore, the book will give a short introduction to the components that are used in the different business model generation approaches such as Osterwalder and Pigneur (2010) and Lindgardt et al. (2009) to reflect about AI implications and growth opportunities.

Osterwalder and Pigneur describe the definition of a business model as how an organization creates, delivers, and captures value (Osterwalder & Pigneur, 2010). Therefore, they created a concept to describe and think through the respective business model. This concept includes nine basic building blocks, which show the logic of how a company intends to make business. According to them, the nine blocks include all the perspectives of the four main areas of a business: customers, offer, infrastructure, and financial viability (Osterwalder & Pigneur, 2010).

Block 1: Customer Segments
It describes the different groups of people or organizations a company aims to reach and serve. Customers comprise the heart of any business model. This segment includes, depending on the business, one or more customer segments with common needs, behaviors, or attributes (Osterwalder & Pigneur, 2010).

Block 2: Value Propositions
It describes the products and services that create value for a specific customer segment. It is the reason why customers will buy or not and it solves a problem or satisfies a need for the customer (Osterwalder & Pigneur, 2010). To define a clear value proposition, Osterwalder created a method called "Value Proposition Canvas" (Osterwalder et al., 2014).

Block 3: Channels
It describes how a company communicates the value proposition with the customer segments (Osterwalder & Pigneur, 2010).

Block 4: Customer Relationships
It describes the types of relationships (personal assistance, self-service, communities, co-creation, etc.) and motivation (acquisition, retention, sales, upselling, etc.) a company form with specific customer segments (Osterwalder & Pigneur, 2010).

Block 5: Revenue Streams
It describes the financial value a company generates from each customer segment: A company must ask itself, for what value is each customer segment truly willing to pay (Osterwalder & Pigneur, 2010)?

Block 6: Key Resources
It describes the assets (physical, financial, intellectual, human, etc.), which are required to run the business model (Osterwalder & Pigneur, 2010).

Block 7: Key Activities
It describes the most important activities that a company must do to run the business model. The key activities are, depending on the business model type, different and often categorized in production, problem solving, platform, etc. (Osterwalder & Pigneur, 2010).

Block 8: Key Partnerships
It describes the required partnerships (strategic alliances, coopetition, joint ventures, buyer-supplier), and suppliers (Osterwalder & Pigneur, 2010).

Block 9: Cost Structure
It describes all costs to run the business model and focuses on either a cost- or a value-driven structure (Osterwalder & Pigneur, 2010).

Lindgardt et al. (2009) describe a business model (BM) based on six components. Overall, the business model is more than a product, service, or technological innovation. It goes apart from single-function strategies, like upgrading the sourcing or sales model.

The value proposition answers the question of what the company offers to whom and reflects explicit decisions for the three sub-elements:

- *Target Segments*: Which customers do the company target and which needs does it address?
- *Product or Service Offering*: What does the company offer to their customers to meet the demand of their needs?
- *Revenue Model*: How does the company get compensated for the offering?

The operating model answers the question of how the company profitably delivers the offering and captures the choices for the three sub-elements:

- *Value Chain*: How does the company arrange to deliver the customer demand? What can the company do inhouse and what is necessary to outsource?
- *Cost Model*: How does the company configure the assets and costs to deliver, based on the value proposition?
- *Organization*: How does the company deploy and develop the employees to sustain and enhance a competitive advantage?

7.5   Implementation Challenges of TAI

Besides following guidance, some challenges for AI system developers and deployers may arise. The implementation of unintentional negative consequences occurs when AI systems are deployed without compliance efforts and without a robust governance (Eitel-Porter, 2021). The TAII Framework helps companies to reduce those unintentional negative consequences by exploring the broader ecosystem and analysis of hidden dependencies. Companies that are more innovative are exposed to a higher risk of implementing unintentional negative consequences. Reasons for this are short development cycles, lack of technical understanding, no established quality assurance, usage of AI outside the original defined context, improper combination of data, and unreported concerns of employees (Eitel-Porter, 2021). Ethical AI systems also require strong governance controls, including process management audit procedures. Implementing (AI) ethics generates costs for AI system development. Aligning the stakeholders, defining a responsible ethics team, considering interdisciplinary perspectives, adapting development, testing concepts, etc. requires additional resources. During the analyses of the TAII Framework stakeholder step, costs and resources can be planned, calculated, and allocated. Ethical considerations may also conflict with commercial interests. The evaluation of the AI system's broader impact, including the ecosystem outside the technical development environment, by using the TAII Framework supports to identify of possible conflicts in an early stage. Finally, the executive management needs to commit and to make the final decisions.

During the second European AI Alliance Assembly, a survey identified the following challenges to AI system deployment: explainability, trust and accuracy, privacy, ethics, predictive accuracy, and transparency (European Commission, 2020b). Research shows that most deployed AI systems are neither transparent nor comprehensible to their users. Rather, they are only interpretable to the engineers used to debug the algorithm (Bhatt et al., 2020). Using the TAII Framework examines the impact on common good and analyses affected stakeholders, including engineers but also people with non-technical background. A successful approach is to extend and split transparency and explainability into two layers: a version on a technical level that is understandable for engineers and another one that is more abstract and understandable for the audience of non-technicians (Bhatt et al., 2020). Transparency exists in many different forms: data, decision-making process, communication, etc. and it limits harm and increases people's trust (Shklovski et al., 2021). Documentation is a key element for all 12 steps of the TAII Framework to achieve transparency and requires both a clear process and a routine "when, what, who and how-to" protocol (Shklovski et al., 2021). Considering the complexity of AI systems, it is hardly possible to achieve total explainability and it is difficult to determine how much explainability is required (Shklovski et al., 2021). The analysis of the broader stakeholders (technical and non-technical) following the TAII Framework creates an understanding of the needs, expectations, and concerns of the affected groups. Based on the stakeholder interviews, legal requirements, risk assessment and common good aspects, AI system developer, and deployer need to decide on the level of explainability.

Values and ethics should be inseparable and intertwined. As all stakeholders should have a common understanding of AI and ethical terms, the author recommends using an aligned definition of specialist terms. Already this first step can be challenging for companies as there are no commonly agreed definitions. One approach is to follow the terms of the European Commission (European Commission AI HLEG, 2019d, 2020). Some initiatives propose AI ethics implementation with a "technical and design" expertise by providing technical concepts for privacy, fairness, etc. (Greene et al., 2019). Technical definitions or explanations for such technological proposals rarely exist (Hagendorff, 2020) and the results of practical implementations are unclear. The TAII Framework supports organizations to start exploring AI ethics from a broader non-technical perspective and makes it possible to deep dive into technical and design implementations in later iterations. AI ethics implementation will be difficult to solve only with a technical approach (Mittelstadt, 2019) and there is no a universal implementation formula, and it cannot be realized by a blueprint. The TAII Framework offers a holistic guidance, and the concrete actions need to be aligned depending on the area of application, context, and used AI technology. Accountability is a central topic for the development of AI systems and data processing, but it comes along with many unanswered questions about the responsibilities of AI system developers (Shklovski et al., 2021; Baker-Brunnbauer, 2021a). As this is a general challenge for all AI system developer and deployer, a common legal framework that specifies the accountability is required. During the pass-through of the TAII Framework all important actions that might influence the accountability should be documented as long as there is no binding legal framework. The risk assessment can be conducted with the two-axis Risk Matrix (Artificial Intelligence Ethics Impact Group AIEI Group, 2020). It defines the risk level on the x-axis by the intensity of possible harm (the number of people that are negatively impacted, the negative impact on society, and the impact on fundamental rights, equality, and social justice).

The y-axis visualizes the dependency of the potentially affected stakeholders on the AI system's actions and decisions by considering control (lower demand for regulation as the AI system operates without human intervention), switchability (possibility of exchanging the AI system and monopolistic dependency to a supplier or producer), and redress (the time needed to understand and correct an unintended AI system outcome). AI systems with the same AI technology can be assigned to different risk levels. Therefore, the application context and area, purpose, training data, and the AI system requirements for the risk assessment need to be included. AI high-risk applications will need to undergo an extra certification to gain the CE marking for distribution and usage within the European Union (European Commission, 2021b). The TAII Framework supports companies in developing trustworthy AI systems before undergoing the conformity assessment.

Without existing legal regulations, the classification as well as the whole AI ethics implementation is not binding but helps companies prepare for upcoming regulatory implementation, managing their product portfolio, and improving their social impact (Baker-Brunnbauer, 2021a). Engineers need to engage openly with internal and external stakeholders on how ethical principles can be implemented. Their organization should support this engineer engagement, expand documentation practices, and should enable support and exchange with public authorities, organizations, and stakeholders (Shklovski et al., 2021). The TAII Framework can be used in stakeholder workshops to introduce the participants to AI ethics, its challenges and to generate an overall understanding of the trustworthy AI implementation requirements. All stakeholders are responsible for sharing their opinions and for making specific proposals for accountable structures. Therefore, organizations need to encourage a culture of open learning and ensure the distribution of responsibility and accountability within the company by implementing standards, assessments, documentation, and testing (Shklovski et al., 2021). Besides the AI system's development life cycle, the Technology Readiness Level (TRL) of the AI technology itself needs to be assessed (Martínez-Plumed et al., 2020). Each of the ethical principles needs to be translated into design and technical requirements that reflect the principles' aim (Anabo et al., 2019; La Fors et al., 2019). This requires a translation in the level of abstraction from principles to micro ethics (Hagendorff, 2020) for reducing abstract norms and generating a Minimum Viable Ethical Product (MVEP) that is useful for people with diverse backgrounds (Jacobs & Huldtgren, 2021). AI system ideation and development workshops using the TAII Framework generate additional value of sustainability and common good along with the value proposition and its technical feature set. The TAII Framework evolves its power by being used in an early stage of the AI system development to include the broader perspectives of involved stakeholders along the value chain, legal requirements, risk consequences, and common good from the beginning. Non-transparent and incomprehensible AI systems will never be socially acceptable because humans will feel controlled by them (Floridi & Taddeo, 2016).

7.6   TAII Framework Canvas

Based on the upcoming regulation of the European Commission of artificial intelligence (AI) systems, organizations and companies that develop, sell, or implement AI systems must develop their AI system accordingly—based on risk classification. This regulation is currently in progress and is known as the "AI Act" (European Commission, 2021d) and "Legal Framework for AI" (European Commission, 2021b) under the name "Trustworthy Artificial Intelligence (TAI)" by working groups of the European Commission. A special focus is on the development of trustworthy and ethically correct functioning AI systems. This means that organizations and companies (AI system developer and deployer) must adapt their AI products and services accordingly or, depending on the risk classification, may not be able to sell or use them within the European Union.

The research study (Baker-Brunnbauer, 2021a) identified a gap in tension between research, development of regulation, and implementation in organizations and companies. The research analyzed the understanding and expectations of managers in AI companies regarding AI ethics and social impacts on their AI development and implementation. The results showed that small and medium-sized enterprises (SMEs) face major challenges and that specific specialist knowledge usually first must be built up. This excessive demand is also because there are no "instructions" or reference projects for any guidance.

Further research led to the development of the TAII Framework (Trustworthy Artificial Intelligence Implementation) (Baker-Brunnbauer, 2021b). This holistic framework offers organizations and companies the first orientation in 12 steps for the implementation of TAI. Since the current approach of the European Commission is very product-oriented and product-regulated, the TAII Framework also includes a strong focus on the analysis of social implications through the whole AI system development process. Thus, within the iterative framework, additional aspects such as the fulfilment of the Sustainable Development Goals (United Nations Sustainable Development Goals, n.d.), UN Human Rights (United Nations Universal Declaration of Human Rights, n.d.), effects on the business model (see Sect. 7.4), supply chain, stakeholders, and certifications are included.

In 2021, the TAII Framework has been presented at various international conferences and events such as Forum Europe, AI4EU and is used in teaching lectures and courses at several international universities such as Ryerson University Canada, NEOMA Business School France, and Mälardalen University Sweden. The TAII Framework is continuously evaluated and adapted to various business cases. The results led to a set of tools to support AI system-developing companies on their way to generate and implement trustworthy AI systems. One elementary workshop tool is the TAII Framework Canvas (Fig. 7.2) which covers the 12 steps of the holistic framework on one side but also generates many possibilities for group interaction between different stakeholders by using a similar approach as the Design Thinking methodology to overcome barriers (Liedtka, 2018).

The following 12 steps of the TAII Framework are included in the canvas:

- *Company Values*: Used for the development of the company's Business Model as well as to shape the AI System Brief Overview.
- *Business Model*: Provides a holistic picture of how the organization creates and captures value.
- *AI System Brief Overview*: Describes the purpose, use case, and the used input data of the AI system.
- *Stakeholder*: Includes all internal and external involved people, groups, departments, companies, organizations, institutions, clusters, etc.
- *Justice*: Considers existing regulations and standards for the specific AI system. Safeguard a high standard of transparency, respect for democratic values, and legitimacy.
- *Risk*: Assessment of the AI system's ethical impact potential harm, and the affected human groups including the unintended results of the AI system.
- *Common Good*: Analyses the dependencies of the 17 Sustainable Development Goals and the Universal Declaration of Human Rights.
- *Ethics*: Generates the core list of ethical requirements. Human agency and oversight; robustness and safety; privacy and data governance; transparency; diversity; non-discrimination and fairness; societal and environmental well-being; and accountability.
- *Translation*: Transfer and translation of the ethical principles and requirements for the AI system's ecosystem.
- *Merge*: Consolidation of the assessed input factors. Definition of the current state, visualization of the dependencies, and planning of the next tasks for improvement.

- *Execution*: Test, implementation, and verification of the results.
- *Certification*: Safety assessment of the AI system based on legal regulations or taken actively into account by the company to provide transparency.

The TAII Framework Canvas can be used in offline workshops by simple paper prints or as a digital template for shared online tools like for example Miro (n.d.). So far, the usage shows an increasing engagement between stakeholders and overcomes the barrier of tensions between groups with different interests. As the adoption of AI systems and their complexity continue to increase and the benefits remain significant (McKinsey Analytics, 2021), the development of trustworthy AI systems, taking into account their social impact and ethical implications, must be a central element in the AI systems development process.

7.7 Conclusion

TAII cannot change a purposefully unethical company strategy and it supports those whose intention is to take self-responsibility for the environment and its living beings. Management needs to generate awareness and implement ethical guidelines within their organization. During the assessments, the focus is on questions that cannot be easily answered. Diverse teams and knowledge plus independent consultancy support the implementation of TAI. The arising questions and outcomes must be aligned with applicable laws and regulations. The implementation of TAI requires a strong long-term company commitment to the development, deployment, and usage of trustworthy AI systems. Tensions between principles and values may arise between the assessment and implementation of trustworthy AI as there is no solution that fits for all AI systems. Stakeholders should analyze the ethical dilemmas with evidence-based reflections and avoid making random decisions. Challenges may arise during the translation and implementation of the multidiscipline topic as values, morals, and ethics are not understandable for AI systems by default and commercial interest may generate tensions. A company's culture and its values as well as their business model will strongly influence the success of trustworthy AI implementations.

(continued)

(continued)

Talk about AI ethics is increasing, and management consulting companies view the topic as a future game changer for companies. Empirical research shows that the practical implementation of AI ethics is lagging and that companies are in the early stage of seeking out guidance. The Trustworthy AI Implementation (TAII) Framework initiates this process and supports management in orienteering and implementing AI ethics within their organization. With the TAII Framework, the management team and its ethics board can explore the systemic dependencies inside and outside their organization.

The Journey Continues
Trustworthy AI and the TAII Framework
Download more information and working material:

References

Aldewereld, H., Dignum, V., & Tan, Y. (2015). Design for values in software development. In J. van den Hoven, P. Vermaas, & I. van de Poel (Eds.), *Handbook of ethics, values, and technological design* (pp. 831–845). Springer. https://doi.org/10.1007/978-94-007-6970-0_26

Anabo, I. F., Elexpuru-Albizuri, I., & Villardón-Gallego, L. (2019). Revisiting the Belmont Report's ethical principles in internet-mediated research: Perspectives from disciplinary associations in the social sciences. *Ethics and Information Technology, 21*, 137–149. https://doi.org/10.1007/s10676-0 18-9495-z

Artificial Intelligence Ethics Impact Group AIEI Group. (2020). *From principles to practice. An interdisciplinary framework to operationalise AI ethics*. Retrieved 9.1.2021 from https://www.ai-ethics-impact.org/resource/blob/1961130c6 db9894ee73aefa489d6249f5ee2b9f/aieig%2D%2D-report%2D%2D-download-hb-data.pdf

Baker-Brunnbauer, J. (2019). *Business model innovation in a paradoxical area of conflict (executive summary)*. https://doi.org/10.13140/RG.2.2.24272.66566

Baker-Brunnbauer, J. (2021a). Management perspective of ethics in artificial intelligence. *AI and Ethics, 1*, 173–181. https://doi.org/10.1007/s43681-020-00022-3

Baker-Brunnbauer, J. (2021b). TAII framework for trustworthy AI systems. *ROBONOMICS: The Journal of the Automated Economy, 2*, 17. https://papers.ssrn.com/sol3/papers.cfm?abstract_id=3914105

Ben-Israel, I., Cerdio, J., Ema, A., Friedman, L., Ienca, M., Mantelero, A., Matania, E., Muller, C., Shiroyama, H., & Vayena, E. (2020). *Towards regulation of AI systems*. Council of Europe Study. Accessed January 9, 2021, from https://rm.coe.int/prems-107320-gbr-2018-compli-cahai-couv-texte-a4-bat-web/1680a0c17a

Bhatt, U., Xiang, A., Sharma, S., Weller, A., Taly, A., Jia, Y., Ghosh, J., Puri, R., Moura, J. M. F., & Eckersley P. (2020). *Explainable machine learning in deployment*. Retrieved from https://arxiv.org/abs/1909.06342v4.

Boddington, P. (2021). AI and moral thinking: How can we live well with machines to enhance our moral agency? *AI and Ethics, 1*, 109–111. https://doi.org/10.1007/s43681-020-00017-0

Braband, J., & Schäbe, H. (2020). On safety assessment of artificial intelligence. *Dependability, 4*, 25–34. https://doi.org/10.21683/1729-2646-2020-20-4-25-34

Brandon, J. (2021). Using unethical data to build a more ethical world. *AI and Ethics, 1*, 101–108. https://doi.org/10.1007/s43681-020-00006-3

Calo, M. R. (2011). Peeping Hals. *Artificial Intelligence, 175*(5-6), 940–941. https://doi.org/10.1016/j.artint.2010.11.025

Cihon, P., Maas, M. M., & Kemp, L. (2020). Fragmentation and the future: Investigating architectures for international AI governance. *Global Policy, 11*(5), 545–556. https://doi.org/10.1111/1758-5899.12890

Clark, T., Osterwalder, A., & Pigneur, Y. (2012). *Business model you*. Wiley.

Council of Europe. (2020a). *Ad hoc committee on artificial intelligence (CAHAI). Feasibility study*. Accessed January 18, 2021, from https://rm.coe.int/cahai-2020-23-final-eng-feasibility-study-/1680a0c6da

Council of Europe. (2020b). *Possible introduction of a mechanism for certifying artificial intelligence tools and services in the sphere of justice and the judiciary*. Accessed December 14, 2020, from https://rm.coe.int/feasability-study-en-cepej-2020-15/1680a0adf4

Cremers, A. B., Englander, A., Gabriel, M., Hecker, D., Mock, M., Poretschkin, M., Rosenzweig, J., Rostalski, F., Sicking, J., Volmer, J., Voosholz, J., Angelika Voss, A., & Wrobel, S. (2019). *Trustworthy use of artificial intelligence. Priorities from a philosophical, ethical, legal, and technological viewpoint as a basis for certification of artificial intelligence*. Accessed January 9, 2021, from https://www.iais.fraunhofer.de/content/dam/iais/KINRW/Whitepaper_Thrustworthy_AI.pdf

Delipetrev, B., Tsinaraki, C., & Kostic, U. (2020). *Historical evolution of artificial intelligence. EUR 30221 EN*. Publications Office of the European Union. https://doi.org/10.2760/801580

Dignum, V. (2019). *HUMANE AI – Toward AI systems that augment and empower humans by understanding us, our society and the world around us*. Retrieved 9.1.2021 from https://www.humane-ai.eu/wp-content/uploads/2019/11/D13-HumaneAI-framework-report.pdf

DKE German Commission for Electrical, Electronic & Information Technologies of DIN and VDE. (2020). *German standardization roadmap on Artificial Intelligence*. Accessed January 9, 2021, from https://www.dke.de/resource/blob/2017010/99bc6d952073ca88f52c0ae4a8c351a8/nr-ki-english%2D%2D-download-data.pdf

Eitel-Porter, R. (2021). Beyond the promise: Implementing ethical AI. *AI and Ethics, 1*, 73–80. https://doi.org/10.1007/s43681-020-00011-6

European Commission. (2018). *Annex to the coordinated plan on artificial intelligence.* Accessed August 5, 2020, from https://ec.europa.eu/newsroom/dae/document.cfm?doc_id=56017

European Commission. (2020a). *White Paper on Artificial Intelligence: A European approach to excellence and trust.* Accessed February 19, 2020, from https://ec.europa.eu/info/publications/white-paper-artificial-intelligence-european-approach-excellence-and-trust_en

European Commission. (2020b). *Second European AI Alliance Assembly.* Accessed January 8, 2021, from https://ec.europa.eu/digital-single-market/en/news/second-european-ai-alliance-assembly

European Commission. (2021a). *Data protection. Rules for the protection of personal data inside and outside the EU.* Accessed February 24, 2021, from https://ec.europa.eu/info/law/law-topic/data-protection_en

European Commission. (2021b). *Regulatory framework proposal on Artificial Intelligence.* Accessed July 8, 2021, from https://digital-strategy.ec.europa.eu/en/policies/regulatory-framework-ai

European Commission. (2021c). *Excellence and trust in artificial intelligence.* Accessed July 8, 2021, from https://ec.europa.eu/info/strategy/priorities-2019-2024/europe-fit-digital-age/excellence-trust-artificial-intelligence_en

European Commission. (2021d). *Artificial Intelligence Act.* Accessed December 5, 2021, from https://eur-lex.europa.eu/legal-content/EN/TXT/?qid=16233 35154975&uri=CELEX%3A52021PC0206

European Commission AI HLEG. (2019a). *Policy and investment recommendations for trustworthy artificial intelligence.* Accessed January 8, 2021, from https://ec.europa.eu/digital-single-market/en/news/policy-and-investment-recommendations-trustworthy-artificial-intelligence

European Commission AI HLEG. (2019b). *Sectoral considerations on the policy and investment recommendations for trustworthy artificial intelligence.* Accessed January 8, 2021, from https://futurium.ec.europa.eu/sites/default/files/2020-07/Sectoral%20Considerations%20On%20The%20Policy%20And%20Investment%20Recommendations%20For%20Trustworthy%20Artificial%20Intelligence_0.pdf

European Commission AI HLEG. (2019c). *Ethics guidelines for trustworthy AI.* Accessed January 8, 2021, from https://ec.europa.eu/newsroom/dae/document.cfm?doc_id=60419

European Commission AI HLEG. (2019d). *A definition of AI: Main capabilities and disciplines.* Accessed January 8th, 2021, from https://ec.europa.eu/newsroom/dae/document.cfm?doc_id=60651

European Commission AI HLEG. (2020). *The Assessment List for Trustworthy Artificial Intelligence (ALTAI) for self assessment.* Accessed January 8, 2021, from https://ec.europa.eu/newsroom/dae/document.cfm?doc_id=68342

European Parliamentary Research Service. (2020). *Artificial intelligence: From ethics to policy.* Accessed January 6, 2021, from https://www.europarl.europa.eu/RegData/etudes/STUD/2020/641507/EPRS_STU(2020)641507_EN.pdf

European Union Agency for Fundamental Rights. (2020). *AI policy initiatives (2016–2020).* Accessed July 4, 2020, from https://fra.europa.eu/en/project/2018/artificial-intelligence-big-data-and-fundamental-rights/ai-policy-initiatives

Fjeld, J., Achten, N., Hilligoss, H., Nagy, A., & Srikumar, M. (2020). Principled artificial intelligence: Mapping consensus in ethical and rights- based approaches to principles for AI. *Berkman Klein Center Research Publication, 1,* 1–39. https://doi.org/10.2139/ssrn.3518482

Floridi, L., & Taddeo, M. (2016). What is data ethics? *Philosophical Transactions of the Royal Society A, 374*(2083), 20160360. https://doi.org/10.1098/rsta.2016.0360

Frankenberger, K., Weiblen, T., Csik, M., & Gassmann, O. (2013). The 4I-framework of business model innovation: A structured view on process phases and challenges. *International Joirnal of Product Development., 18,* 249–273. https://doi.org/10.1504/IJPD.2013.055012

Floridi, L., Cowls, J., Beltrametti, M., Chatila, R., Chazerand, P., Dignum, V., Luetge, C., Madelin, R., Pagallo, U., Rossi, F., Schafer, B., Valcke, P., & Vayena, E. (2018). AI4People—An ethical framework for a good AI Society: Opportunities. *Risks, Principles, and Recommendations, Minds & Machines, 28,* 689–707. https://doi.org/10.1007/s11023-018-9482-5

Gassmann, O., Frankenberger, K., & Csik, M. (2014). *The business model navigator.* Pearson Education Limited.

Girotra, K., & Netessine, S. (2014). *Four paths to business model innovation.* Accessed February 24, 2019, from https://hbr.org/2014/07/four-paths-to-business-model-innovation

Greene, D., Hoffmann, A. L., & Stark, L. (2019). Better, nicer, clearer, fairer: A critical assessment of the movement for ethical artificial intelligence and machine learning. In *Proceedings of the 52nd Hawaii International Conference on System Sciences (HICSS, 2019)* (pp. 2122–2131). https://doi.org/10.24251/HICSS.2019.258

Hagendorff, T. (2020). The ethics of AI ethics: An evaluation of guidelines. *Minds and Machines, 30,* 99–120. https://doi.org/10.1007/s11023-020-09517-8

Hickok, M. (2021). Lessons learned from AI ethics principles for future actions. *AI and Ethics, 1,* 41–47. https://doi.org/10.1007/s43681-020-00008-1

KI Strategie Deutschland. (2020). *Artificial Intelligence Strategy of the German Federal Government.* Accessed March 2, 2021, from https://www.ki-strategie-deutschland.de/files/downloads/Fortschreibung_KI-Strategie_engl.pdf

Jacobs, N., & Huldtgren, A. (2021). Why value sensitive design needs ethical commitments. *Ethics and Information Technology, 23,* 23–26. https://doi.org/10.1007/s10676-018-9467-3

Jobin, A., Ienca, M., & Vayena, E. (2019). The global landscape of AI ethics guidelines. *Nature Machine Intelligence, 1*(9), 389–399. https://doi.org/10.1038/s42256-019-0088-2

Johnson, M. W., Christensen, C. M., & Kagermann, H. (2008). *Reinventing your business model.* Accessed August 9, 2019, from https://hbr.org/2008/12/reinventing-your-business-model

Krafft, T. D., & Zweig, K. A. (2019). *Transparenz und Nachvollziehbarkeit algorithmenbasierter Entscheidungsprozesse | Ein Regulierungsvorschlag.* Accessed February 17, 2021, from https://www.vzbv.de/sites/default/files/downloads/2019/05/02/19-01-22_zweig_krafft_transparenz_adm-neu.pdf

La Fors, K., Custers, B., & Keymolen, E. (2019). Reassessing values for emerging big data technologies: Integrating design-based and application-based approaches. *Ethics and Information Technology, 21,* 209–226. https://doi.org/10.1007/s10676-019-09503-4

Lauer, D. (2021). You cannot have AI ethics without ethics. *AI and Ethics, 1,* 21–25. https://doi.org/10.1007/s43681-020-00013-4

Liedtka, J. (2018). *Why design thinking works.* Harvard Business Review. Accessed December 15, 2021, from https://hbr.org/2018/09/why-design-thinking-works

Lindgardt, Z., Reeves, M., Stalk, G., & Deimler, M. S. (2009). *Business model innovation.* Accessed March 12, 2019, from https://www.bcg.com/documents/file36456.pdf

Mantelero, A. (2018). AI and big data: A blueprint for human rights, social and ethical impact assessment. *Computer Law & Security Review, 34*(4), 754–772. https://doi.org/10.1016/j.clsr.2018.05.017

Martínez-Plumed, F., Gómez, E., & Hernández-Orallo, J. (2020). *AI Watch, assessing technology readiness levels for artificial intelligence.* Joint Research Center European Commission. https://doi.org/10.2760/15025

Matzler, K., Bailom, F., Friedrich von den Eichen, S., & Anschober, M. (2016). *Wie Sie Ihr Unternehmen digital auf das digitale Zeitalter Disruption vorbereiten.* Vahlen.

McKinsey Analytics. (2021). *The state of AI in 2021.* Accessed December 15, 2021, from https://www.mckinsey.com/business-functions/mckinsey-analytics/our-insights/global-survey-the-state-of-ai-in-202

Miro. (n.d.). *The visual collaboration platform for every team.* Accessed December 15, 2021, from https://miro.com

Mittelstadt, B. (2019). Principles alone cannot guarantee ethical AI. *Nature Machine Intelligence, 1,* 501–507. https://doi.org/10.1038/s42256-019-0114-4

Morley, J., Floridi, L., Kinsey, L., & Elhalal, A. (2019). *A typology of AI ethics tools, methods and research to translate principles into practices.* Accessed October 10, 2020, from https://aiforsocialgood.github.io/neurips2019/accepted/track2/pdfs/26_aisg_neurips2019.pdf

Ryan, M., & Stahl, B. C. (2021). Artificial intelligence ethics guidelines for developers and users: Clarifying their content and normative implications. *Journal of Information, Communication and Ethics in Society, 19*(1), 61–86. https://doi.org/10.1108/JICES-12-2019-0138

Osterwalder, A., & Pigneur, Y. (2010). *Business model generation.* Wiley.

Osterwalder, A., Pigneur, Y., Bernarda, G., & Smith, A. (2014). *Value proposition design.* Wiley.

Ovans, A. (2015). *What is a business model?* Accessed August 9, 2019, from https://hbr.org/2015/01/what-is-a-business-model

Shafer, S. M., Smith, H. J., & Linder, J. C. (2005). The power of business models. *Business Horizons, 48,* 199–207. https://doi.org/10.1016/j.bushor.2004.10.014

Shklovski, I., ANE, Data Ethics ThinkDoTank, & IEEE. (2021). *Addressing ethical dilemmas in AI: Listening to engineers.* Accessed January 27, 2021, from https://futurium.ec.europa.eu/sites/default/files/2021-01/Addressing%20Ethical%20Dilemmas%20in%20AI%20%E2%80%93%20Listening%20to%20the%20Engineers.pdf

Systemic Society. Deutscher Verband für systemische Forschung, Therapie, Supervision und Beratung e.V. (n.d.). *Systemische Methoden.* Accessed August 21, 2021, from https://systemische-gesellschaft.de/systemischer-ansatz/methoden

Thiebes, S., Lins, S., & Sunyaev, A. (2021). Trustworthy artificial intelligence. *Electronic Markets, 31*, 447–464. https://doi.org/10.1007/s12525-020-00441-4

Twomey, P., & Martin, K. (2020). *A step to implementing the G20 principles on artificial intelligence: Ensuring data aggregators and AI firms operate in the interests of data subjects.* Accessed January 8, 2021, from https://www.g20-insights.org/wp-content/uploads/2020/04/g20-principles-artificial-intelligence-data-aggregators-ai-firms-1586167851.pdf

Umbrello, S., & van de Poel, I. (2021). Mapping value sensitive design onto AI for social good principles. *AI and Ethics.* https://doi.org/10.1007/s43681-021-00038-3.

United Nations Sustainable Development Goals. (n.d.). Accessed January 9, 2021, from https://sdgs.un.org

United Nations Universal Declaration of Human Rights. (n.d.). Accessed January 9, 2021, from https://www.un.org/en/universal-declaration-human-rights

Vakkuri, V., Kemell, K.-K., Kultanen, J., Siponen, M., & Abrahamsson, P. (2019). *Ethically aligned design of autonomous systems: Industry viewpoint and an empirical study.* Retrieved from http://arxiv.org/abs/1906.07946

Wachter, S. (2019). Data protection in the age of big data. *Nature Electronics, 2*, 6–7. https://doi.org/10.1038/s41928-018-0193-y

World Economic Forum. (2019). *AI Governance: A Holistic Approach to Implement Ethics into AI.* Accessed April 28, 2020, from https://www.weforum.org/whitepapers/ai-governance-a-holistic-approach-to-implement-ethics-into-ai